Karl Riedling

Ellipsometry for Industrial Applications

Springer-Verlag Wien New York

The Author

Dr. Karl Riedling, born 1948 in Vienna, Austria, works as an Assistant Professor at the Institut für Allgemeine Elektrotechnik und Elektronik of the Technical University Vienna. The present work has been sub-mitted as part of the author's habilitation thesis.

With 43 Figures

ISBN-13: 978-3-211-82040-7   e-ISBN-13: 978-3-7091-8961-0
DOI: 10.1007/978-3-7091-8961-0

Preface

During the past decade, ellipsometry acquired increasing importance in industrial areas, particularly in the technology of microelectronic devices. As a contact–less, non–destructive optical surface analysis technique which allows the accurate measurement of the thicknesses of films in the sub–micrometer region and which permits a qualitative and quantitative assessment of the composition and/or structure of the sample under investigation, ellipsometry is specially suited for the contamination–sensitive manufacturing environment and the delicate structures of electronic devices. Ellipsometers which were specially designed for the demands of the semiconductor and thin film industry and which featured a continuously increasing degree of automation of the measurement and its analysis became commercially available from the mid–1970s on; most of these instruments, however, and the soft-ware required for the evaluation of their measurements, are supplied as "black boxes" to the industrial user who frequently has to rely on them without any detailed knowledge of the accuracy of measured data. Particularly if samples become more complex, measurement and evaluation errors made unknowingly may add up, and measured data may finally become meaningless.

This work endeavors to present some considerations on the accuracy limitations of the two basic types of ellipsometers which are currently in use in the industrial environment, and on the limitations of the technique itself. Design guidelines for ellipsometers are given which are based upon these error estimations. The possibilities and potential problems of ellipsometric measurements made *in situ*, during the deposition or removal of films on the surface of a sample, are discussed. Furthermore, a short overview over the most important applications of ellipsometry in the research on semiconductor and thin film devices is provided which may indicate the vast potential of this technique which extends, indeed, far beyond the simple measurements of the thick-nesses of dielectric layers which constitute today's main industrial use of ellipsometry.

Vienna, November 1987   *Karl Riedling*

Acknowledgements

This work was supported by the Fonds zur Förderung der wissenschaftlichen Forschung, Vienna, Austria. Printing was made possible by a grant of the Österreichische Forschungsgemeinschaft.

Some of the material in this volume has been previously published in one of the following papers, and is reproduced here with the permission of Elsevier Sequoia, Lausanne, and Springer, Berlin – Heidelberg – New York – Tokyo, respectively:

Riedling, K.: Evaluation of adjustment data for simple ellipsometers. Thin Solid Films, 61, 335 (1979).

Riedling, K.: Error effects in the ellipsometric investigation of thin films. Thin Solid Films, 75, 355 (1981).

Riedling, K.: Dynamische *in situ*-Ellipsometrie für Grundlagenuntersuchungen und Prozesskontrolle. Fresenius Z. Anal. Chem., 319, 706 (1984).

Riedling, K.: Accuracy of digital Fourier transformation detection systems for high speed rotating analyzer ellipsometers. Thin Solid Films, 155, 151 (1987).

Contents

List of Illustrations

List of Tables

1. Basics of Ellipsometry

1.1 Physics

Ellipsometry is a technique for the contact–less and non–destructive optical characterization of surfaces [1],[2],[3],[4]. It is based on the fact that a monochromatic electromagnetic wave changes its state of polarization if it strikes non–perpendicularly the interface between two dielectric media. In general, any arbitrary monochromatic transversal wave can be considered composed of two orthogonal coherent waves with a fixed phase relation, e.g., of two linearly polarized waves whose electric field vectors lie within two perpendicular planes (Fig. 1). The field vector E_{res} which results from a vector addition of the two components E_x and E_y lies within a plane if and only if the two constituent waves are in phase, or out of phase by a multiple of π (which corresponds to an inversion of one of them); the resulting wave is linearly polarized in this case (Fig. 1 (a)). Otherwise, the field vector performs a screw–like motion around the direction of its propagation (the z–direction); its projection onto a plane perpendicular to its direction of propagation describes, in general, an ellipse (Fig. 1 (b)). This ellipse degenerates to a line if the phase difference between the two constituent waves is a multiple of π; it becomes a circle if their amplitudes are equal, and the phase difference between them is $\pi/2$ (90°).

The orthogonal components of an arbitrary wave need not necessarily be linearly polarized. Two coherent circularly polarized waves whose field vectors rotate clockwise and counterclockwise, respectively, fulfill the same purpose. Indeed, any representation of a wave can be chosen which suits the problem in mind best, if only the constituents are orthogonal waves.

The reflection of an electromagnetic wave at and its transmission through the interface between two homogeneous media is described by Fresnel's equations (Fig. 2) which can be derived from Maxwell's equations, taking into consideration the continuity of the tangential field components at the interface.

(a)

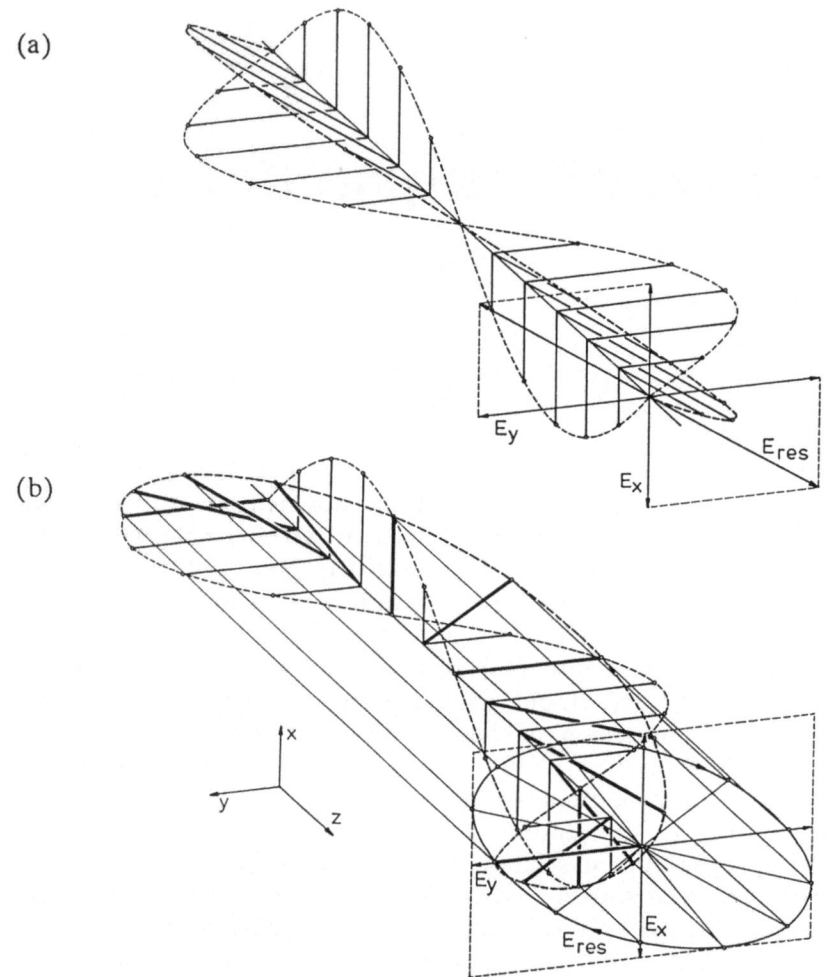

(b)

Fig. 1:   Linear (a) and elliptic (b) polarization of an electromagnetic
          wave.

For the analysis of a system like the one shown in Fig. 2, it is ex-
pedient to assume the constituent waves to be polarized in parallel to
the plane of incidence (index "p" in Fig. 2), and perpendicular to it
(index "s"), respectively. (The plane of incidence is defined by the
propagation directions of the incident and reflected beams, or, equivalently, by the propagation direction of the incident beam and the
normal to the interface.) In general, the "p" and the "s" components
are reflected and transmitted to a different degree according to Fresnel's equations, resulting in polarization dependent reflection and

transmission coefficients r_{01} and t_{01}, respectively. The state of polarization of the wave is changed in this case; since this change is a function of the optical parameters of the entire system, it can be used to determine some of them, provided all other parameters are known.

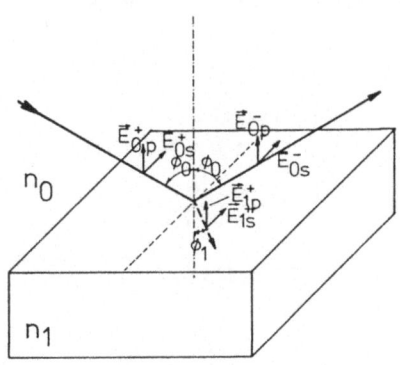

FRESNEL'S EQUATIONS:

$$r_{01P} = \frac{E_{0P}^-}{E_{0P}^+} = \frac{n_0\cos\phi_1 - n_1\cos\phi_0}{n_0\cos\phi_1 + n_1\cos\phi_0}$$

$$t_{01P} = \frac{E_{1P}^+}{E_{0P}^+} = \frac{2n_0\cos\phi_0}{n_0\cos\phi_1 + n_1\cos\phi_0}$$

$$r_{01S} = \frac{E_{0S}^-}{E_{0S}^+} = \frac{n_0\cos\phi_0 - n_1\cos\phi_1}{n_0\cos\phi_0 + n_1\cos\phi_1}$$

$$t_{01S} = \frac{E_{1S}^+}{E_{0S}^+} = \frac{2n_0\cos\phi_0}{n_0\cos\phi_0 + n_1\cos\phi_1}$$

with: $n_0\sin\phi_0 = n_1\sin\phi_1$

Fig. 2: Reflection at the interface between two homogeneous media.

If absorbing materials and/or more complex structures than the one in Fig. 2 are considered, it is convenient to use complex notation for the description of the wave components. Fresnel's equations contain only real components for an interface between two absorption-free, quasi-infinite dielectric media; the resulting reflection coefficients are real in this case, too, which means that there is no phase shift (except possibly by an integer multiple of π) between the incident and reflected or transmitted beams. Such an interface affects only the ratio of the amplitudes of the reflected or transmitted wave components but not their phase relation; a linearly polarized wave is thus returned from it as a linearly polarized wave, although, in general, with a different azimuth. In contrast, a phase shift between the two orthogonal components of the wave is introduced either if one of the two media is absorbing, or if multiple reflections occur in a thin layer of a dielectric medium embedded between two arbitrary media (Fig. 3). In the first case, Fresnel's equations contain complex refractive indices and result therefore in complex reflection and transmission coefficients; in the second, the resulting total reflection and transmission coefficients R and T as defined in Fig. 3 are, in general, complex, due to the superposition of an infinite number of reflected and transmitted waves whose relative phases differ because of the different distances they travelled. (There are different total reflection and transmission coefficients for the "p" and the "s" wave components since r_{ik} and t_{ik} are different for both orthogonal components. The "p" and "s" indices were omitted in Fig. 3 for reasons of brevity.) In either instance, a phase shift is introduced between the incident and reflected beams; the phase shifts for the two orthogonal components of an arbitrary wave are most likely different. In general, a linearly polarized incident wave will therefore be reflected or transmitted with elliptic polarization in the two cases mentioned.

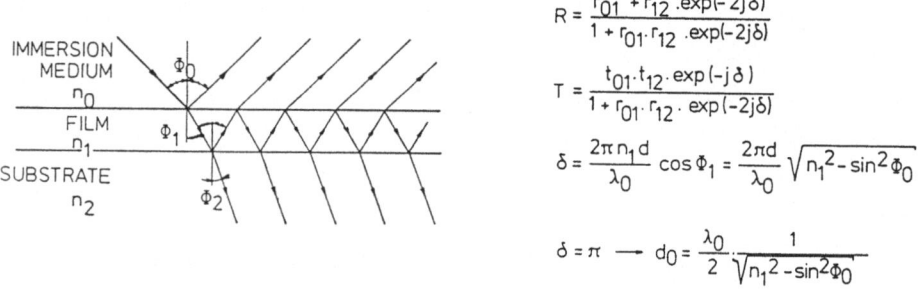

$$R = \frac{r_{01} + r_{12} \cdot \exp(-2j\delta)}{1 + r_{01} \cdot r_{12} \cdot \exp(-2j\delta)}$$

$$T = \frac{t_{01} \cdot t_{12} \cdot \exp(-j\delta)}{1 + r_{01} \cdot r_{12} \cdot \exp(-2j\delta)}$$

$$\delta = \frac{2\pi n_1 d}{\lambda_0} \cos\Phi_1 = \frac{2\pi d}{\lambda_0} \sqrt{n_1^2 - \sin^2\Phi_0}$$

$$\delta = \pi \longrightarrow d_0 = \frac{\lambda_0}{2} \cdot \frac{1}{\sqrt{n_1^2 - \sin^2\Phi_0}}$$

Fig. 3:   Reflection at and transmission through a thin dielectric film.

Ellipsometry entails the quantitative analysis of such changes of the polarization, usually of an electromagnetic wave within or near the visible part of the electromagnetic spectrum, and the interpretation of the measured data with regard to the optical properties of the sample under investigation. Although experimental problems increase if wavelengths outside the visible spectral range are used, ellipsometers operating, for example, with the wavelength of a CO_2 laser (10.4 μm) [5] or even in the microwave region [6] have been reported. Most practical implementations of ellipsometry utilize the reflected beam only, which permits their application to arbitrary – transparent or opaque – substrates. Accordingly, the following considerations will be restricted to reflection ellipsometry in or next to the visible spectral range.

Normally, sample parameters from an ellipsometric measurement are evaluated in two steps: First, a "complex reflection coefficient" ρ of the surface under investigation is calculated from the actually measured data, e.g., from the azimuth angles of polarizing prisms at which extinction of the light finally falling on a photodetector occurs, or from the directly determined azimuth and ellipticity of a polarization ellipse. This complex reflection coefficient is defined as the quotient of the actual (and, in fact, also complex) reflection coefficients for light polarized in parallel (R_p) and perpendicular (R_s) to the plane of light incidence, respectively, which, in turn, are defined as the ratios of the complex amplitudes of the reflected and incident wave components (compare Fig. 2). Since ρ holds only the ratio of the intensities of both components and their relative phase difference, it can be determined without measuring absolute intensities and phases, which facilitates the measurement and eliminates the influence of possible fluctuations of the absolute intensity and phase of the original probe beam. Usually, ρ is transformed into the "ellipsometric angles" Ψ and Δ according to the relation

$$\rho = R_p/R_s = \tan \Psi \cdot \exp (j \cdot \Delta),$$

which permits to map the entire complex plane to the area $0 \leq \Psi \leq \pi/2$, $0 \leq \Delta \leq 2\pi$. (In most cases, Ψ and Δ are referred to in terms of degrees, though, rather than radians.) The two scalar ellipsometric angles (or, the real and imaginary parts of ρ) obtained from one ellipsometric measurement permit the calculation of two independent scalar parameters of the sample; all other optically relevant parameters of the sample and of the measurement system have to be known [1],[7]. The separation of the measurement evaluation into the calculation of ρ or Ψ and Δ, and the subsequent analysis of these data makes the actual determination of the sample parameters independent from the experimental setup used, which is particularly important since a variety of ellipsometric techniques and instruments have been reported which may differ significantly, considering the instrument parameters primarily measured [1].

The most straightforward case in the evaluation of sample data is the determination of the refractive index and the absorption coefficient of a potentially absorbing homogeneous sample whose thickness is large enough to prohibit multiple reflections at its top and bottom surfaces (or where, at least, the beam reflected from the bottom surface is offset far enough from the beam reflected at the actually investigated top surface to be excluded from the measurement by the limited optical aperture of the detector). Obviously, such a measurement can be used for the characterization of a bulk sample, particularly if measurements are taken for an entire range of wavelengths. Since the results are affected by the presence of thin surface layers like native oxide films, special precautions are usually required such as cleavage of the samples within an ultra-high vacuum environment immediately prior to the measurement. It is, in general, not possible to investigate semiconductor substrates without such safeguards. Even sub-monolayers of oxide films on substrates like silicon can easily be detected ellipsometrically; substrate parameters obtained by neglecting the presence of such films are hardly more than meaningless [8].

The parameters determined most frequently in technical applications of ellipsometry, especially in microelectronics, are the thickness and the refractive index of a thin dielectric or slightly absorbing film on a substrate with known properties. The refractive index of the film renders valuable information about the film composition and/or structure [7],[9]; particularly for films in the thickness region of a few hundred nanometers or less, ellipsometry is the only technique which permits a reliable thickness measurement. It can be shown that – within a certain thickness range – each pair of film thickness and refractive index values can be unambiguously and reversibly mapped to a pair of Ψ/Δ values. (Due to the inherent periodicity of the interference phenomena which lead to the complex behavior of the total reflection coefficients for thin films, this effect is periodic with the film thickness; its period is the thickness d_0 of an "ellipsometric order" as defined in Fig. 3.) Graphs similar to the one schematically shown in Fig. 4 were widely used for the evaluation of ellipsometric film thickness measurements prior to the advent of inexpensive and sufficiently powerful microcomputers.

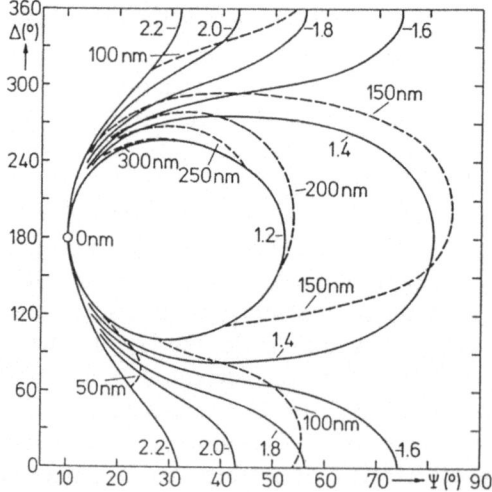

Fig. 4: Ψ–Δ diagram for the estimation of the thicknesses and refractive indices of thin films on a silicon substrate. (Solid lines: constant refractive index, broken lines: constant film thickness.)

In general, any two parameters even of arbitrarily complex samples may (theoretically) be determined from one ellipsometric measurement if all other optically relevant parameters are known, for example the thickness and the refractive index of one particular film in a multi-layer structure, or the parameters of absorbing films [7],[10],[11]. Even very inhomogeneous films may be investigated ellipsometrically if it is possible to monitor their growth or removal *in situ*; the considerations and techniques applied to this kind of analysis will be discussed in more detail in chapter 1.4. Such complex measurements require a very thorough knowledge of the sample under investigation, and other physical or chemical analysis methods may have to be applied in addition in order to permit a correct interpretation of the ellipsometric data. However, the results rendered by such a combined analysis may by far exceed the sum of those which could be obtained with each method separately. In contrast to many other analytic techniques, ellipsometry derives the major part of its flexibility and performance from the proper interpretation of the measured data.

1.2 Instrumentation

Two basic approaches can be distinguished which permit to determine the complex reflection coefficient of a sample, namely, null ellipsometry, and photometric ellipsometry [1]. Null ellipsometry comprises all techniques which are based on measuring the state of polarization indirectly by adjusting the azimuths of polarizing prisms and other instrument parameters for an extinction of the light falling on a photodetector. Photometric ellipsometry, in contrast, determines the state of polarization of the probe beam more directly by suitable measurements of its intensity.

1.2.1 Null Ellipsometry

Because of its simpler measurement evaluation algorithms and due to a smaller influence of experimental imperfections, null ellipsometry was originally used almost exclusively; with the exception of instruments offered by one manufacturer only, all currently commercially available equipment for the industrial environment is still based on this technique.

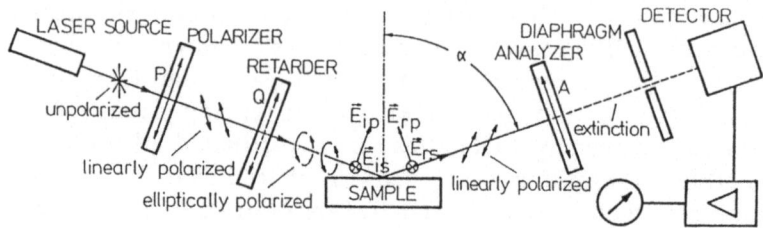

Fig. 5:   Principle of a PRSA null ellipsometer.

In the most common PRSA (Polarizer – Retarder – Sample – Analyzer) null ellipsometer configuration (Fig. 5), an elliptically polarized light beam is incident on the sample. Its ellipticity and azimuth are chosen in such a way that the reflection at the sample surface just converts the beam to linear polarization. The ellipticity of the incident beam is controlled by a polarizer with a variable azimuth which produces a linearly polarized wave. This wave subsequently passes through a retarder, a quarter–wave plate of a birefringent material. Depending on the relative positions of the transversal axes of the polarizer and the retarder, the resulting beam can be adjusted to any state of polarization between linear and circular: The difference of the azimuth angles of the polarizer and the retarder defines the ellipticity of the beam, while the azimuth of the retarder determines the azimuth of the polarization ellipse. A linearly polarized beam is reflected from the sample if and only if the ellipticity of the incident light is exactly compensated for by the reflection at the sample. Only in this case, the reflected beam can be totally extinguished by a second polarizer, the analyzer. A measurement is done by alternating adjustments of the polarizer and analyzer azimuths for minimum intensity (the retarder is normally kept at a fixed azimuth); its results are two independent sets of polarizer and analyzer azimuth readings for which extinction occurs.

Compared to photometric ellipsometry, this approach has several important advantages: The complex reflection coefficient can very easily be obtained from the polarizer and analyzer azimuths at extinction if certain retarder settings are used, and if the optical components are assumed to be ideal [12]. Light source intensity fluctuations, residual polarization, polarizer imperfections, and environment light scattered into the optical path are of minor influence on the result. Sample conditions which prohibit a reasonably accurate ellipsometric measurement, such as surface roughness or lateral inhomogeneities, usually result in a flat and less pronounced intensity minimum, thus warning an experienced operator of potentially erroneous measurements. The experimental technique and the processing of the measured data can be refined relatively easily in order to improve the intrinsic accuracy or sensitivity of the instrument [13].

The most severe disadvantage of this technique, though, is that it requires repeated alternating adjustments of the polarizer and analyzer azimuths for each measurement, which results in relatively long data acquisition times: Even a skilled operator needs in the order of one minute per measurement on a manually operated instrument. The speed of automated null ellipsometers is limited by the fact that the directions of the movements of the polarizer and analyzer prisms have to be reversed many times before a valid result is available. Systems using mechanically tuned polarizing prisms (as opposed to faster but more expensive electro–optical devices) are impeded by the mechanical inertia of the prism stages which confines the achievable speeds to relatively small values. These speed limitations prohibit the application of null ellipsometry particularly for the *in situ* monitoring of dynamic processes: the strongly non–linear relation between the measured data and the sample parameters may result under certain conditions in very drastic and fast changes of the complex reflection coefficient (and, thus, of the measured null azimuths) although the sample may have been subject to a minor modification only. Furthermore, the azimuths of the polarizer and analyzer have to be measured with a resolution and accuracy in the order of hundredths of a degree to guarantee a reasonable accuracy, which tightens the requirements for the mechanical assembly and for the azimuth measurement.

Chapter 3.2.1 will be devoted in more detail to two particular aspects of null ellipsometry, namely, to the consequences of an imperfect alignment of such instruments and of possibly non–ideal optical components [8], and to their numerical calibration [14].

1.2.2 Photometric Ellipsometry

Speed and accuracy considerations generally favor the second major ellipsometric technique, photometric ellipsometry [1]. Among the numerous variations of this principle (e.g., "polarization modulated ellipsometry" [15]), a technique called "rotating analyzer ellipsometry" [16],[17],[18] is used most extensively (Fig. 6): A linearly polarized light beam (or, in some varieties, a beam with known elliptic or circular polarization) is incident on the sample, from which it is reflected, in general, elliptically polarized. The reflected beam passes through a second polarizer, the analyzer, which rotates at a constant and, within the limitations of the mechanical design, arbitrarily high angular speed. The intensity behind the analyzer is measured with a suitable detector. Since only that component of the reflected beam may pass the analyzer which is currently polarized in parallel to its transmission plane, the intensity of the light falling on the detector will be modulated with twice the rotation frequency of the analyzer. If the beam incident on the analyzer were linearly polarized, a sine-square function would result for the intensity detected, with one maximum and one zero-intensity minimum per half rotation of the analyzer. Evidently, no intensity fluctuations would ensue at all if the reflected beam were circularly polarized. Elliptically polarized light results in sinusoidal fluctuations similar to those seen for linearly polarized light but with a smaller modulation depth, i.e., with a smaller amplitude variation. The modulation depth is therefore a function of the ellipticity of the reflected light beam, and the phase angle of the modulation relative to the zero azimuth of the analyzer represents the azimuth of the polarization ellipse (compare Fig. 6).

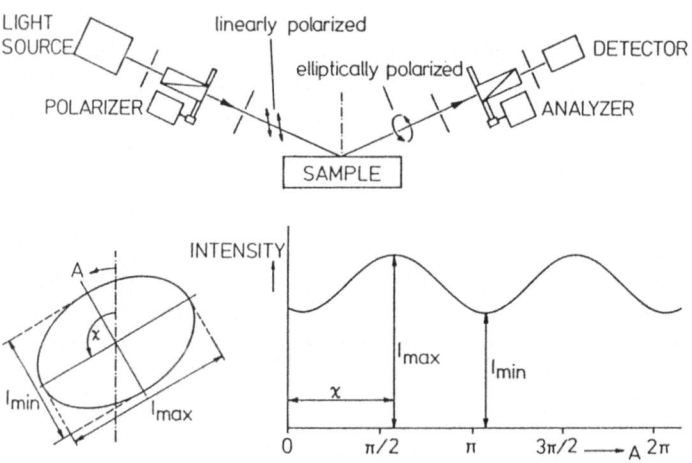

Fig. 6:   Principle of a rotating analyzer ellipsometer.

A variety of the rotating analyzer technique operates at the principal angle of incidence. The principal angle is defined as the angle of incidence at which a phase shift of $\pi/2$ occurs between the two orthogonal components of the wave which are polarized in parallel and perpendicular to the plane of incidence, respectively [1]. For absorption-free dielectric materials, the principal angle coincides with the Brewster angle (where the component polarized in parallel to the plane of incidence is extinguished and where it changes its sign); absorption of the substrate introduces a shift between the principal and the Brewster angles which lies typically between fractions of a degree and a few degrees, depending on the substrate absorption coefficient. If the angle of incidence is equal to the principal angle, the reflected beam is elliptically polarized with either the major or the minor axis of the polarization ellipse in parallel to the plane of incidence. Its ellipticity can be tuned by modifying the azimuth of the polarizer; the polarizer azimuth is exactly equal to Ψ if the reflected beam is just circularly polarized (compare eqs. (35) – (37) in chapter 3.2.2.4). The measurement proper is done by adjusting the instrument for these conditions. This setup results in an improved signal–to–noise ratio, particularly in spectroscopic applications [19],[20].

Modulation depth and phase angle can be derived from the measured data by analog methods (e.g., with a lock–in amplifier), or, more suitably for the further data processing, by a digital fast Fourier analysis of the measured intensity. The results of this Fourier analysis permit to calculate the complex reflection coefficient of the sample which can be further processed exactly as for a null ellipsometer.

The major advantages of rotating analyzer ellipsometers lie in their inherently higher speeds, and in their increased accuracies. Since several hundreds or even thousands of intensity samples constitute one single measurement, the effect of random errors and noise is significantly reduced. The absolute azimuthal position of the analyzer need not be known more accurately than within fractions of a degree (its reproducibility must be better, though); samples of the light intensity at the photodetector may be taken at analyzer azimuth intervals ranging from fractions of a degree to tens of degrees. In contrast to null ellipsometry, a retarder is not necessarily required. Omitting the retarder has not only the advantage that the relatively large errors usually caused by commercial wave plates [8] can be avoided; it simplifies the evaluation of the measured data particularly in spectroscopic measurements since the properties of a wave plate are to a certain degree wavelength dependent.

The photometric approach raises some significant problems unknown to null ellipsometry, though: Imperfections of the polarizers [21], stray light falling into the optical path, and fluctuations of the intensity of the light source may cause severe errors which have to be compensated for by suitable methods. The response of the photodetector has to be strictly linear to avoid the generation of harmonics. On the other hand, the speed requirements imposed on the detector and the amplifier circuitry are not too stringent since the signal frequency is exactly twice the rotation frequency of the analyzer prism. Even for the most elaborate setups with mechanical components, this frequency will hardly exceed a few hundreds of Hertz. Phase shifts and delays caused by the electronic system, and azimuthal alignment errors of the sample and the analyzer can be easily compensated for in a relatively simple calibration procedure prior to each measurement [18].

The potentially shorter data acquisition time and the simplified retarder-less construction of a rotating analyzer ellipsometer favor this design for spectroscopic investigations where a large number of data points have to be acquired within a reasonable period of time in order to obtain a spectrum of the sample parameters, and for fast *in situ* measurements. Still, up to date there are no instruments commercially available whose data acquisition times are less than in the order of seconds; a commercial instrument which is based on the rotating analyzer approach and which is designed for the *in situ* monitoring of microelectronic gas plasma processes operates with an analyzer rotation speed of less than 1 rpm [22]. Some considerations for the construction of an instrument which will be by about two orders of magnitude faster are presented in Appendix A; the accuracy of rotating analyzer ellipsometers will be discussed in chapter 3.2.2.

1.3 Developments in Ellipsometry

Although the principles of ellipsometry date back to the early 19th century when Fresnel formulated the equations which are known under his name [23], this technique was not widely used until the beginning 1960s. The evaluation of ellipsometric measurements entails the solution of relatively complicated complex transcendental equations which can be solved analytically only in a few special cases but require numerical approaches in most instances. An ellipsometric analysis was, therefore, not practical prior to the availability of sufficiently powerful data processing equipment. Only gradually, ellipsometry was introduced into an industrial area like microelectronic technology, first, with nomographs similar to the one shown in Fig. 4 for the measurement evaluation which had been generated by the instruments' manufacturers on a mainframe computer [12]; by the end of the 1970s, self-contained units were introduced on the market which feature, in general, an automated measurement and internal data processing, based on a microcomputer [22],[24],[25]. However, most of these instruments, and the computer programs supplied with them for measurement evaluation, are tailored to large-scale standard applications like the measurement of the thicknesses and refractive indices of homogeneous thin dielectric films which were deposited directly on a semiconductor substrate with known optical properties (like single crystal silicon).

Indeed, the evolution of ellipsometry split into two different branches during the last decade: While the instruments designed for industrial applications sacrificed flexibility and, to a certain degree, accuracy for the ease of operation, the physicists re-discovered the inherent potential of ellipsometry either for the determination of very complex material parameters, or for the *in situ* monitoring of dynamic physical or chemical processes without interference.

The endeavors in material characterization led to the development of spectroscopic ellipsometry, i.e., ellipsometry which uses not only one wavelength of monochromatic light but takes a number of measurements at different wavelengths within an interesting range, mostly, within the entire visible spectrum and parts of the near infra-red and ultra-violet [3]. Spectroscopic ellipsometry thus determines not only one set of optical parameters, e.g., the refractive and absorption indices of an absorbing sample at one particular wavelength, but entire spectra of these parameters which may contain valuable information about the energy band structure of the specimen [26] or may permit the detection of structural defects like dislocations or voids on the surface of a single crystal sample [27], or interface films in complex multi-layer systems [28],[29],[30],[31].

The second scientific aspect of ellipsometry, its capability of monitoring dynamic processes *in situ*, proved increasingly important in widely different areas such as surface physics [32], chemistry [33], semiconductor [34],[35],[36],[37], and plasma technology [38],[39],[40],[41]. In general, each process which affects a surface to some degree, either by explicit deposition or removal of films, or by modification of the substrate surface, can be monitored by ellipsometric methods, provided the light beam used does not cause an undue interference with the process under observation. This is particularly important in areas where the samples have to be embedded in gaseous or liquid media which prohibit the use of analytic tools other than optical, and where it is essential that the reaction dynamics be monitored.

1.4 Static and Dynamic Ellipsometry

The conventional applications of ellipsometry comprise the analysis of completely prepared samples, for example, for the refractive index and the absorption of a homogeneous substrate, or for the thickness and the refractive index of a thin film. The validity of data obtained from such a single "static" measurement is, however, limited by the fact that a relatively simple structure of the sample – either a homogeneous piece of bulk material without any surface layers, or a homogeneous film with abrupt interfaces to a homogeneous substrate and to the immersion medium – must be taken for granted in order to permit an evaluation of the measurements. Aside from very few special cases (like thermally grown oxide on silicon), it is very likely that many samples do not entirely comply with the restrictions imposed by the simple evaluation models. Interface layers like native oxides or inhomogeneities over the thickness of a film may or may not be present, depending on the particular preparation technique. They may adversely influence the quality of a film, and it should be the commission of a film characterization technique to discover them reliably. A single ellipsometric measurement will, however, not only fail to detect them but be falsified to a more or less unacceptable degree if their presence is neglected [8].

Vertical inhomogeneities or interface layers can easily be accounted for if the ellipsometric angles Ψ and Δ are calculated from known sample parameters, i.e., in a procedure opposite to the interpretation of measured ellipsometric angles. Any inhomogeneous film can be considered approximated by a stack of sufficiently thin homogeneous layers with suitably graded refractive indices n (Fig. 7).

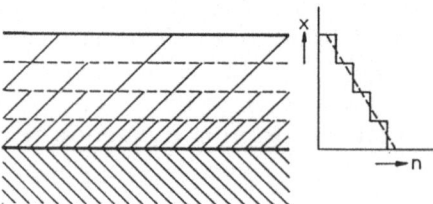

Fig. 7:   Approximation of an inhomogeneous film by a stack of quasi-homogeneous films.

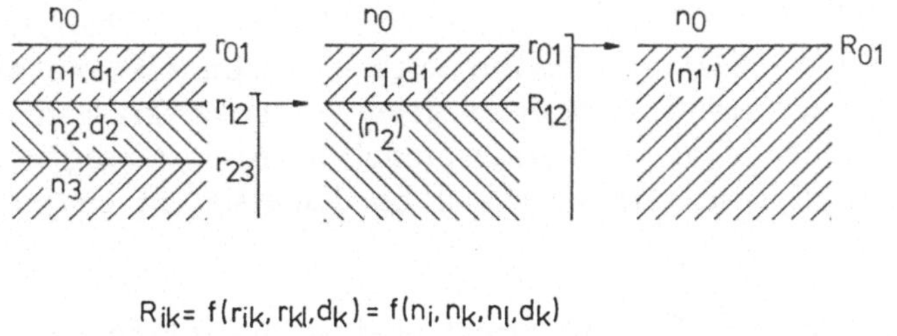

$$R_{ik} = f(r_{ik}, r_{kl}, d_k) = f(n_i, n_k, n_l, d_k)$$

Fig. 8:   Determination of the reflection coefficients in a multi–layer system.

The total reflection coefficients R_p and R_s (for the wave components polarized in parallel and perpendicular to the plane of incidence, respectively) of such a stack, and hence Ψ and Δ, can be determined according to Fig. 8:   In the first step, the total reflection coefficients R_{12p} and R_{12s} are defined similar to the approach used for calculating the total reflection coefficient of a single homogeneous film (compare Fig. 3); the refractive index of the next film on top of it (n_1 in Fig. 8) is used as an "immersion medium". (The indices "p" and "s" were omitted in Figs. 3 and 8 for reasons of simplicity.) R_{12p} and R_{12s} describe the reflection at the interface between the media 1 and 2, no matter from which actually underlying structure it resulted. The substrate and the film next to the substrate can therefore be considered combined into a fictive substrate with the – generally complex – refractive index n_2', and the above procedure can be repeated for an arbitrary number of layers [42]. (Actually, it is not n_2' which enters into the further evaluation but the two reflection coefficients R_{12p} and R_{12s} for the two orthogonal wave components. The significance of this fact will be discussed in chapter 3.3.) Under the assumption of a given structure of a film, Ψ and Δ can thus be determined, and calculated data for different models can be compared to measured values. This does not make too much sense, though, for a single measurement because many different models may result in the same or almost the same ellipsometric angles. In general, the comparison of a single "static" measurement to data derived from a model does not give conclusive evidence about the actual film structure.

The above considerations can also be applied to devise algorithms for the direct evaluation of measured data for inhomogeneous films if the structure of the inhomogeneity (e.g., the dependence of the refractive index on the distance from the surface of the substrate) is exactly known. Frequently, this is not the case altogether: The models used for describing inhomogeneous films are usually very simple (e.g., assuming a linear dependence of the refractive index [7] or of the composition [8] of a film on the distance from the substrate surface) in order to allow their application with a reasonable expenditure of computing power; practical samples need not necessarily comply with them. The influence of inaccurate modelling of inhomogeneous films on the calculated film thickness will be discussed in chapter 3.3 for some special cases.

Furthermore, methods for the direct analysis of inhomogeneous films are hampered by the fact that only two unknown parameters can be derived from a single ellipsometric measurement, which is usually not sufficient for the analysis of inhomogeneous films which are defined by at least three parameters (the refractive indices at the top and the bottom of the film, and the film thickness). Additional information can be derived from multiple ellipsometric measurements of the same sample either under various angles of incidence [43], or with several wavelengths [44]. Still, the analysis of such multiple measurements is very complicated, and the results obtained are frequently ambiguous.

While a single "static" measurement results in exactly one point in the Ψ-Δ plane (compare Fig. 4), which permits to calculate exactly two sample parameters, this is no more true if series of measurements are taken during the growth or etching of a film. Each single measurement can be used to determine the parameters of the currently grown (or removed) "differential" layer, based on the fictive substrate parameters which result from the layers beneath. If it is possible to monitor the growth of an arbitrarily inhomogeneous film *in situ*, measurements can be taken at intervals which are considered short enough to guarantee quasi-homogeneity of the layer deposited since the last measurement was taken. Under favorable conditions, the thickness and refractive index of the last "differential" layer can be calculated, using the parameters of the previously deposited and evaluated layers. An inhomogeneous film is thus "broken down" into a stack of quasi-homogeneous films whose thicknesses and refractive indices can be derived separately. Analogously, data obtained from an etching process monitored *in situ* can be evaluated. It should be mentioned, though, that this technique can hardly be applied to films with rapid changes of their refractive indices: It is necessary to determine the thickness and the refractive index of each "differential" layer in order to have these data available for the analysis of the next measurement. It is, however, hardly possible to calculate both values with sufficient accuracy for films whose thickness is less than 40 or 50 nanometers [8]; the investigated films should, therefore, be reasonably homogeneous within any 50 nm "slice".

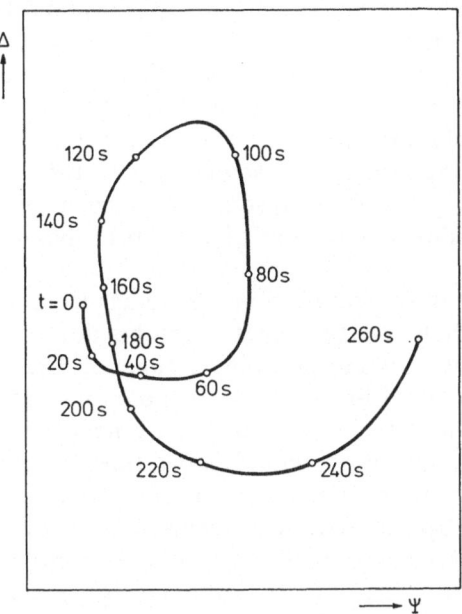

Fig. 9:   Graph of the ellipsometric angles Ψ and Δ as a function of
time obtained from an *in situ* measurement (schematically).

Since any dynamic *in situ* measurement results in a "fingerprint" of
the sample in the Ψ–Δ plane (as shown schematically in Fig. 9), rather
than a single point [43], the "fingerprint" can be matched to Ψ–Δ function graphs obtained from various models, thus permitting the selection
of the best fitting model, even if accuracy limitations prohibit the
direct analysis of each "differential" layer. (Such investigations have,
for example, been reported on the nucleation of layers at an elec-
trode–electrolyte interface where even conclusions were drawn to the
shape of droplets formed at nucleation sites [33].)

Aside from the obvious applications of *in situ* ellipsometry for the
analysis of genuine multi–layer films, or of films whose compositions
or microstructures vary gradually, resulting in a gradual variation of
the refractive index, the *in situ* approach is also suitable for the
investigation of films with a micro–rough surface. Surface roughness
corresponds to a zone with a reduced refractive index since it can be
regarded as a mixture of the solid film material and the immersion
medium which is most likely air or vacuum. If the assumption is
justified that the surface roughness of a film does not change significantly during its growth, the growing film can be interpreted as a

zone of homogeneous film material with increasing thickness which is capped by a disturbed surface layer with (presumably) invariant properties. Similarly, the *in situ* technique allows the investigation of films with non-uniformly distributed pinholes, cracks, or voids. It should be noted, however, that lateral structures must be small compared to the wavelength used for the measurement in order to result in an average refractive index; structures in the order of the wavelength or greater prohibit ellipsometric measurements on principle; they simply cause de-polarization of the probe beam.

The main constraint for reliable *in situ* measurements is the speed of the instrument: each measurement must be done within a time short enough to rule out any measurable change of the sample parameters during the data acquisition. It is, in general, not practical or even impossible to halt a film deposition or etching process for the duration of a measurement; doing so would most likely result in film discontinuities and other adverse effects on the process. For example, the resolution of an ellipsometric measurement is in the order of fractions of a monolayer for very thin silicon dioxide or nitride films on silicon [8]; for a growth rate of, say, 10 nanometers per minute and a maximum permitted thickness change of 0.1 nanometers during a measurement, the maximum permitted data acquisition time is 0.6 seconds. The data acquisition speed of the only currently commercially available *in situ* instrument [22] with a minimum of 1 second per measurement is probably no more sufficient in this regime.

2. Ellipsometry in Microelectronic Technology

2.1 Semiconductor Substrates and Films

Although the industrial use of ellipsometry is essentially limited to the analysis of insulating non–absorbing films, research applications of this technique within the field of microelectronic technology concentrate on the investigation of semiconducting substrates and layers, and the modifications they undergo during various technological processes. In most cases, spectroscopic methods are applied.

In general, ellipsometry is not extremely suited for the characterization of bulk semiconductor material with a low doping density; the detection limit for bulk impurities is in the order of 10^{19} cm^{-3} [5], [26]. Resolution is better, though, in cases where doped or undoped damage layers were formed on the surface of a semiconductor single crystal, e.g., by sputter etching, or by ion implantation. Lattice damage caused by r.f. sputter etching of silicon, and the effects of subsequent annealing cycles, can be investigated by spectroscopic measurements; the ellipsometrically determined changes of the refractive index can be correlated, e.g., to the leakage current level of Schottky diodes built on the etched surfaces [45]. Ion implanted surfaces prior to annealing are not only damaged but contain a relatively high concentration of the implanted impurities (although the effects observed experimentally were, in general, attributed to the damage rather than to the doping); ellipsometric investigations were reported on implanted silicon [46],[47] and GaAs [44] samples. The hypothesis that not the doping density but primarily the lattice damage determines the changes of the refractive index is corroborated by the fact that arsenic doses as low as 10^{12} cm^{-2} caused a measurable effect whereas the detection level for a boron implantation was at a dose of at least 10^{15} cm^{-2} [48]. It is, in general, not possible to obtain quantitative results from

ellipsometric measurements for the thickness of the damaged layer or
the type of damage; the measurements give, however, conclusive qualitative evidence of the presence and degree of surface damage, and of
the lattice restoration achieved with subsequent annealing treatments.

Ellipsometry has also been applied extensively and successfully to the
investigation of epitaxial and polycrystalline films and vertical semiconductor structures [34],[35],[36],[37]. In general, the thickness and
composition of these films can be analyzed quantitatively. *In situ*
ellipsometry has been used with success for the control of the growth
of superlattice structures with molecular beam epitaxy (MBE) or metalorganic chemical vapor deposition (MOCVD) techniques [49]. An optical method like ellipsometry is particularly suited in the latter am-
bience where the standard techniques used in the ultra-high vacuum
environment of MBE (e.g., electron diffraction) fail.

2.2 Insulating Films

Most applications of ellipsometry in the technology of microelectronic
devices comprise the investigation of thin insulating films, usually on
semiconductor substrates. For a variety of industrially used substrate-
film systems, ellipsometry became one of the standard analytic techniques [25],[31]. Aside from thermally grown or anodic oxides [29],
[50], these films are usually deposited by plasma enhanced oxidation
[28],[39] or deposition, or by chemical vapor deposition (CVD) [40]
techniques, or by methods like evaporation or sputtering. These processes are generally industrially established although great efforts are
still required to fully understand the dynamics of the mechanisms involved in some of them [51].

Problems common to all film deposition techniques – which can not be
discussed here in detail due to their variety – are the proper monitoring and control of the film quality and thickness, and the analysis of
possible vertical and lateral structures within the films. While there
are many feasible analytic techniques for completely prepared films
which range from optical and electrical measurements to microanalytic
methods like electron microscopy, SIMS (Secondary Ion Mass Spec-
trometry), X-ray topography, or Auger electron spectroscopy, any
advanced process control approach demands some kind of *in situ* investigation. The technique used must, however, be insensitive to interferences caused by the high r.f. or d.c. fields, high temperatures,
aggressive atmosphere, and adverse pressure conditions which are

typical for most film deposition processes [52]. Furthermore, the measurement itself must not cause any interference with the process under observation. Both demands can be met by a suitable optical technique: A light beam is hardly hampered by the above mentioned conditions, and it is not likely to interfere with the process if the wavelength used is chosen such that no or no essential absorption and therefore energy transfer takes place within the monitored system. The results obtained from *in situ* process monitoring can be used in two ways: They permit to apply closed–loop control to the process, thus allowing to grow films reproducibly with exactly defined characteristics, and they can render valuable information about the process itself, either about the growth dynamics of the films, or about the possible formation of interface layers between the substrate and the films which may affect the operation of devices built with them. Especially for non–plasma film deposition processes, such studies have been reported extensively [53],[54],[55],[56].

2.3 Etching Processes

The various forms of plasma etching processes, including sputter etching, ion, and ion beam etching, and the reactive varieties of these techniques, gained increasing importance in the technology of semiconductor devices during the last decade, essentially because of their advantages, compared to the conventional wet chemical etching processes, in the preparation of micron and sub–micron structures required for the production of very large scale integrated (VLSI) circuits [57],[58],[59].

Similar requirements for an *in situ* monitoring as discussed above arise in plasma etching processes. The reproducibility and yield of an etching process can be significantly improved by an appropriate end point detection method which permits to terminate the etching exactly when the entire layer which has to be etched has been removed but before the substrate or structures under the film have been unduly affected [38]. This is particularly necessary for non–reactive plasma etching processes some of which have only a limited selectivity range, i.e., their etch rates do not differ significantly for different materials. In contrast to a properly designed wet chemical process, and in contrast to many reactive plasma etching processes, etching does not stop automatically in this case when all material to be removed has been etched.

On principle, plasma etching can also be used as an analytic tool: Films of unknown composition and/or structure can be gradually removed, and repeated measurements can be taken of the residual film, advantageously, *in situ*. The evaluation of such experiments is analogous to the analysis of measurements taken during an *in situ* analysis of a film being deposited, only in reversed order. Obviously, such an etching analysis can also be applied to films which cannot be monitored during their growth.

3. Error Effects in Ellipsometric Investigations

Ellipsometric measurements are, like any experiment, subject to various error effects which may distort their results to a more or less critical degree. In general, the following sources of measurement errors can be identified:

(1) Random errors: They may be caused by an imperfect search for the intensity minimum in null ellipsometry, and by various noise sources in photometric approaches. In general, the only way to reduce random errors is taking a number of measurements, and applying statistical procedures to the results. Genuine random errors may be augmented by the effects of imperfections of the instrument which are not known accurately enough to be compensated for. Since these errors are systematic by nature, they cannot be reduced by statistical methods. Still, their effect on the measured data is comparable to the impact of random measurement errors (since it does not matter for a single measurement whether a measured azimuth angle was incorrect because of an erroneous reading or because of a bad alignment of the instrument); if these alignment errors are in the same order of magnitude as the actual random errors, it seems justified to subsume them in the considerations of random errors. These effects will be addressed in chapter 3.1.

(2) Equipment imperfections and alignment errors: The sources and the influence of such error effects depend obviously on the basic type and operation mode of the instrument. Errors may be caused by imperfections of optical components [21], bad mechanical alignment [14], stray light and parasitic beams, noise induced by the light source, noise and distortions caused by the electronic system of the instrument, and quantization noise generated by digital processing of the analog photodetector output [60]. Some of the imperfections of components and alignment can be entirely compensated for by numerical calibration [14] or by alignment procedures which precede the measurement proper [17]. These equip-

ment–based error mechanisms, and algorithms for their compensation, will be discussed in more detail in chapter 3.2.

(3) Sample data errors: The evaluation of an ellipsometric measurement requires the accurate knowledge of all optically relevant parameters of the sample except those two which are to be calculated from the measured data. Frequently, some of the sample parameters are not known accurately enough, which results in errors of the calculated parameters. Such errors may propagate through the analysis of a complex sample if, for example, the ellipsometrically determined parameters of a substrate are used for an evaluation of the data of a subsequently deposited thin film; they may become critical if many evaluation steps have to based on the preceding ones, like in *in situ* ellipsometry. The influence of the uncertainties of the sample structure on the accuracy of an ellipsometric measurement will be reviewed in chapter 3.3.

3.1 Random Measurement Errors

The influence of random error effects on an actual ellipsometric measurement was estimated by means of a computer simulation with a measurement evaluation program which is based on the McCrackin algorithms [7] but has been improved and expanded considerably, compared to the McCrackin program [1]. "Measured" data, either of the ellipsometric angles Ψ and Δ, or of the polarizer and analyzer extinction azimuth angles P and A, were generated while varying system parameters according to miscellaneous error sources. Subsequently, "sample" parameters were calculated from the simulated "measurements" under the assumption of a perfect (or perfectly compensated) alignment and of perfect optical components of the instrument. Although the computer program used was basically developed for the evaluation of measurements done with null ellipsometers, and the simulated "measurements" were performed on a fictive null ellipsometer accordingly, this does, in general, not restrict the validity of the results presented here for other setups. (For example, different effects may cause an error of Ψ and Δ in a null ellipsometer and in a photometric instrument, respectively, but a certain error of the ellipsometric angles will

---

1 Copies of the Fortran source code of this measurement data reduction program, "ELLIPS", and of the pertinent documentation are available from the author.

have the same influence on the parameters of the sample derived from them in either case.) The simulations were essentially based on the data of the system silicon – SiO_2 – air; the simulated data were solved for the thickness and – if possible – for the refractive index of the films. Unless otherwise noted, the following parameters were used in the simulations:

Table 1: Standard data for the computer simulations.

Ellipsometer:	PRSA null ellipsometer
Wavelength:	$\lambda = 632.8$ nm
Angle of incidence:	$\alpha = 70°$
Retarder:	ideal quarter–wave plate:
Transmittance ratio:	$T_c = 1$
Phase shift:	$\Delta_c = 90°$
Azimuth:	$Q = -45°$
Immersion medium:	air ($n_m = 1$)
Film medium:	SiO_2 ($n_f = 1.45$)
	Si_3N_4 ($n_f = 2.00$)
Substrate medium:	silicon ($n_s = 3.85 - 0.02j$)

Interpreting the measured ellipsometric data for the thickness and the refractive index of a film requires the solution of a complex transcendental equation. This equation can be solved analytically only if the refractive index of the film is known. Since only one parameter, namely, the film thickness, is calculated in this case as the solution of a complex equation, the resulting value is generally complex. The thickness of a film must be a real quantity, though; the imaginary part of the result can therefore be regarded as an error term which must disappear if the correct system parameters are used. Numerical algorithms may consequently be applied which vary the refractive index used for the calculation of the film thickness until the imaginary part of the result becomes equal to zero [1],[7].

This standard procedure can, however, not be applied to the measured data in all instances: The dependence of Ψ and Δ on the thickness and the refractive index of a film is strongly non–linear. Although each pair of film thickness and refractive index values corresponds to exactly one point in the Ψ–Δ plane (compare Fig. 4), these points are packed so densely in some regions that even minor errors of the ellipsometric angles may cause significant deviations of the calculated film parameters, or may even result in combinations of Ψ and Δ which do not correspond at all to a system with meaningful physical data.

These areas are next to the limits of an ellipsometric order; they affect ultra–thin films, and films whose thickness is slightly below or, due to the periodicity of the ellipsometric angles as a function of film thickness, slightly above a thickness multiple (i.e., an integer multiple of the period d_0 as defined in Fig. 3). For SiO_2 films on silicon and the set of parameters listed in Table 1, the thickness ranges affected are less than about 40 nm, 250 – 330 nm, and so on. A solution for film thickness <u>and</u> refractive index is no more practical in each of these regimes; film thickness values ought therefore to be obtained with a fixed film refractive index value (which should, of course, be estimated as accurately as possible). This approach will, in general, result in a non–zero imaginary part of the calculated thickness. It was found that, in the vicinity of a thickness multiple, this imaginary part is affected much stronger by the value of the angle of incidence which is used for data evaluation than by a modification of the re- fractive index of the film. It appears therefore reasonable to attrib- ute the imaginary part of the calculated film thickness to a deviation of the actual angle of incidence from the value used in the data reduction procedure rather than to an error of the film refractive index. Accordingly, an algorithm was devised and implemented in the measurement evaluation software which minimizes the imaginary part of the calculated thickness by varying the angle of incidence value in the film thickness computation algorithm. The validity of a particular measurement can be checked easily by comparing the calculated angle of incidence to the corresponding instrument specifications. A mea- surement can be regarded reliable if the deviation between both values can be accounted for by wedge errors of the sample and known align- ment errors of the instrument. In general, all error analyses within the critical ranges were done with this routine; otherwise, the simu- lated data were solved for film thickness <u>and</u> refractive index.

For the estimation of random error effects, a 95% confidence interval (according to the McCrackin program [7]) of ±0.2° for the polarizer and analyzer settings was assumed, which corresponds to confidence limits $\delta\Psi$ of 0.2°, and $\delta\Delta$, of 0.4°, respectively. These values may seem relatively large; however, they were experimentally found to be typical for simple manually operated null ellipsometers with vernier scales for the azimuths. (A quantitative estimation of the uncertainties of the measured ellipsometric angles and the results derived from them, based on known uncertainties of the various parameters of the measurement, is given in the literature for the special case of ultra–thin films [61].) In any case, the results obtained for the above confidence ranges can easily be scaled to match smaller or greater errors since they are sufficiently small to permit linearization. The ellipsometric angles Ψ and Δ for a film with a given thickness and refractive index were adjusted with their respective confidence limits following an algorithm devised by McCrackin [7], according to which the measured ellipsometric angles Ψ_o and Δ_o determine the center, and the confidence limits $\delta\Psi$ and $\delta\Delta$, the axes of an "error ellipse" in the Ψ–Δ plane within which the actual values of Ψ and Δ lie with a 95% probability (Fig. 10).

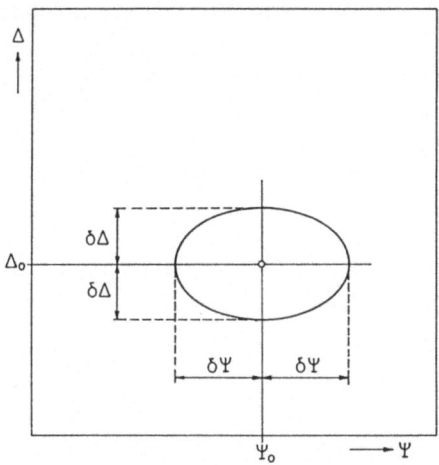

Fig. 10: Random errors: Definition of the "error ellipse" in the Ψ–Δ plane.

The results of this random error analysis are shown in Fig. 11 for a silicon dioxide film on silicon over the first ellipsometric order ($0 \leq d \leq 286.5$ nm). For a given actual film thickness d_a, the relative film thickness error x and the calculated refractive index $(n_f)_c$ of the film will lie with a probability of 95% between the curves shown. The solid lines apply to an evaluation of d <u>and</u> n_f, whereas the broken lines indicate the confidence limits of a calculation of d only with a fixed (correct) value for n_f. The relative error of the film thickness x is defined as

$$x = \frac{d_c - d_a}{d_a} \tag{1}$$

where d_c and d_a are the calculated and actual film thickness values, respectively.

Three thickness ranges can be distinguished in Fig. 11: In the region of ultra–thin films ($d_a < 40 - 50$ nm; range A), and next to the end of the first ellipsometric order ($d_a > 240 - 250$ nm; range C), a calcula–

tion of the film thickness <u>and</u> the refractive index is not possible because parts of the error ellipse lie outside the ranges corresponding to real film thickness and refractive index values. The relative thickness error x increases steeply if the film thickness approaches zero; the absolute thickness error for films with near-zero thickness is slightly less than ± 0.2 nm; it increases slowly (less than proportional to the actual film thickness) with increasing film thickness (compare Fig. 13 (c) in chapter 3.2.1.1, and Fig. 35 (c) in chapter 3.3.1).

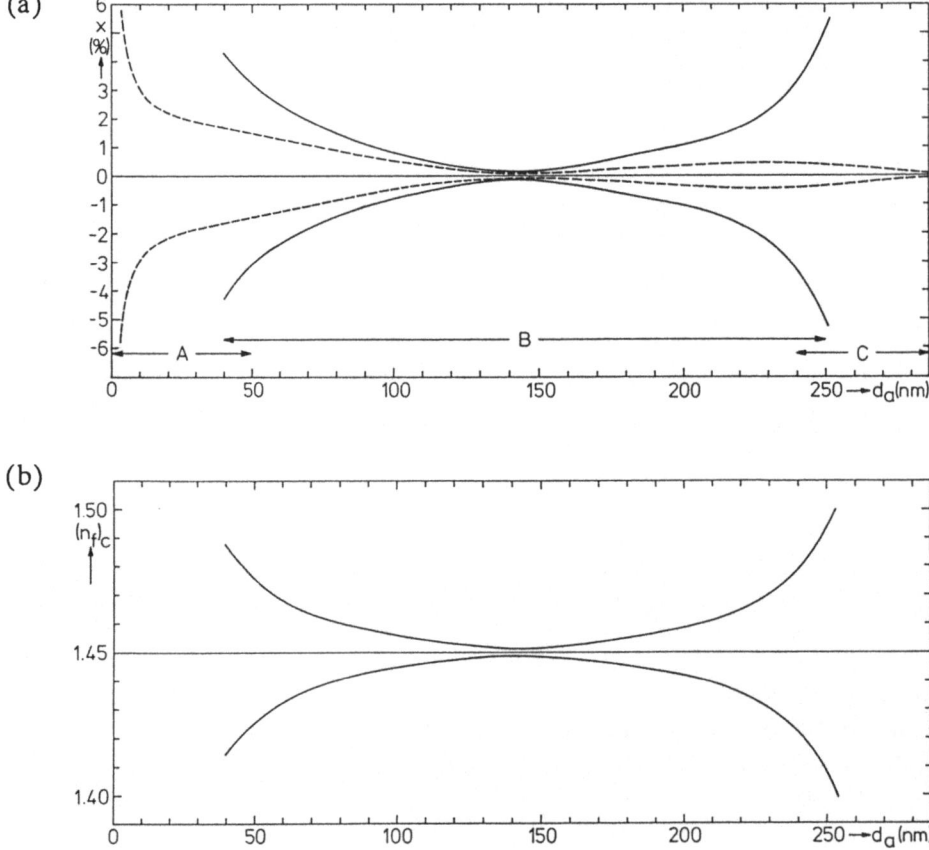

Fig. 11:   Random errors: 95% confidence limits of the relative error x of the calculated thickness (a) and of the calculated refractive index $(n_f)_c$ (b) of an SiO_2 film on silicon over the first ellipsometric order for uncertainties $\delta\Psi = 0.2°$ and $\delta\Delta = 0.4°$. (Solid lines: evaluation for $(n_f)_c$ <u>and</u> d_c; broken lines: for d_c only with n_f kept at 1.45.)

In range B, either measurement evaluation approach – solution for d only, or for d <u>and</u> n_f – is feasible; the errors obtained from the latter are obviously greater. Optimum accuracy can be achieved if the film thickness lies exactly in the center of an ellipsometric order. While the error of the approach which is based on a fixed n_f value (broken lines in Fig. 11(a)) remains moderate for greater film thicknesses, this does not apply to the results of the d/n_f algorithm whose errors increase again drastically (solid lines). This can be attributed to the fact that the calculated film thickness is a function of the refractive index of the film and therefore affected by its errors. The course of x over d_a reflects, therefore, the course of $(n_f)_c$ over d_a in this case.

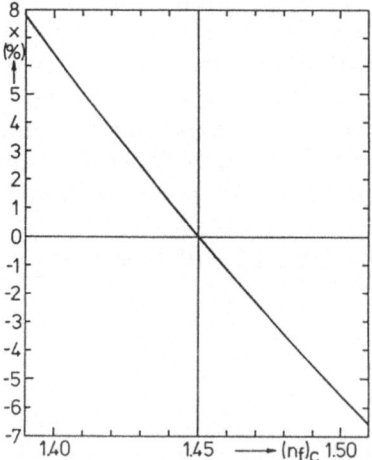

Fig. 12: Relative error x of the thickness period d_0 corresponding to one ellipsometric order as a function of the film refractive index $(n_f)_c$ for an actual index $(n_f)_a = 1.45$.

The calculated thickness d_c of films in a higher than the first ellipsometric order is influenced very strongly by the refractive index of the film which is used for the computation. This is true because the thickness period d_0 is a function of n_f (compare Fig. 3); it is hardly affected by any other error mechanism. The relative error of d_0 depends overproportionately on the error of n_f; Fig. 12 delineates this dependence for a film with the actual refractive index $(n_f)_a = 1.45$ whose thickness is calculated with $(n_f)_c$. The error of d_0 predominates in films which are several ellipsometric orders thick; their total thickness error can therefore be estimated from the data given in Fig. 12. Hence, an accurate determination of the refractive index of such films is an important prerequisite for their correct analysis.

3.2 Instrumentation Error Effects

3.2.1 Null Ellipsometers

3.2.1.1 Error Sources

Null ellipsometers of the PRSA (Polarizer – Retarder – Sample – Analyzer) type were practically the only setup which was commonly used in research as well as an industrial environment prior to the advent of powerful data processing equipment. This was due to the advantages of this technique outlined in chapter 1.2, and particularly because the conversion of the actually measured data – two independent sets of polarizer and analyzer azimuth angles – into the ellipsometric angles Ψ and Δ can easily be done for these instruments by a simple linear combination of the measured azimuths, under the assumption that the retarder is an ideal quarter–wave plate, with its fast axis mounted at $-45°$ relative to the plane of incidence (compare Fig. 5). Under these premises, the following relations will hold [12]:

$$P_2 = P_1 + 90°$$
$$A_2 = 180° - A_1 \tag{2}$$

P_i and A_i denote the i–th set of polarizer and analyzer azimuth angles for extinction, measured counterclockwise from the plane of incidence. From each of the two independent sets, ellipsometric angles Ψ_i and Δ_i can be calculated which result in the same complex reflection coefficient ρ:

$$\Delta_1 = 2 \cdot P_1 + 90° \qquad \qquad \Delta_2 = 2 \cdot P_2 + 90°$$
$$\Psi_1 = A_1 \qquad \qquad \Psi_2 = A_2 \tag{3}$$
$$\rho_1 = \tan \Psi_1 \cdot \exp(j \cdot \Delta_1) \qquad \qquad \rho_2 = \tan \Psi_2 \cdot \exp(j \cdot \Delta_2)$$

$$\rho_1 = \rho_2 = \rho \tag{4}$$

These simple equations are no more valid in any one of the following cases:

(a) The polarizer and/or analyzer are azimuthally misaligned with respect to the plane of incidence of the probe beam.

(b) The retarder is not ideal, i.e., its phase delay Δ_c between the fast and the slow beams differs from 90°, or the ratio T_c of transmittances along its fast and slow axes differs from unity. Unfortunately, the phase delay of a retarder depends on its thickness and therefore on production tolerances; commonly, T_c is not exactly equal to 1 either.

(c) The retarder azimuth angle Q is not exactly –45°.

(d) The surface of the sample is not adjusted perpendicular to the plane of light incidence, which may be due to a wedge shape of the sample, or due to bad instrument alignment. If such a tilt angle is small enough, it has an effect similar to a corresponding azimuthal misalignment of the entire instrument. The tilt angle β can be calculated from two independent pairs of P–A azimuth readings for minimum intensity with a linearized complex approach; the real part of the result can be considered as the actual tilt angle which can be subtracted from all azimuths prior to any further evaluation steps in order to compensate for the "misalignment". The imaginary part of the result is an error term whose magnitude gives some indication whether the deviations from eq. (4) were actually caused by a tilt error (in which case it is close to zero) or by other effects [7].

Different ρ values will result for either set of extinction angles if eqs. (3) are used under non–ideal conditions. The resulting errors can be partly compensated for by using either the average of the two Ψ and Δ [12], or ρ [7] values. (The two approaches are no more equivalent if there are differences between the two sets of Ψ and Δ.) A substantial improvement of the accuracy of the calculated ellipsometric angles can only be achieved if the exact transcendental complex relation linking P and A with Ψ and Δ [1],[7] is used instead of eqs. (3) with the correct parameters of the retarder, and if the magnitude of possible misalignments is known. Averaging the complex reflection coefficients obtained with the exact approach for the two sets of extinction angles may reduce the influence of possible residual errors, and further improve the accuracy by about one order of magnitude:

$$\rho = \tan \Psi \cdot \exp(j \cdot \Delta) = (\rho_1 + \rho_2)/2 \qquad (5)$$

with

$$\rho_i = \tan A_i \cdot \frac{\tan Q + \rho_c \cdot \tan (P_i - Q)}{\rho_c \cdot \tan Q \cdot \tan (P_i - Q) - 1} \qquad i = 1,2 \qquad (6)$$

and

$$\rho_c = T_c \cdot \exp(-j \cdot \Delta_c) \qquad\qquad (7)$$

In order to estimate the influence of the potential error effects on the ellipsometric angles Ψ and Δ computed from the measured data, $\Psi-\Delta$ pairs were calculated for various film thickness values while one of the parameters listed in Table 2 was varied within the range indicated, whereas all other parameters were kept at their ideal values.

Table 2: Apparatus alignment errors used in the simulations.

Azimuthal alignment errors:	
Polarizer:	$\delta P = \pm 0.5°$
Analyzer:	$\delta A = \pm 0.5°$
Retarder:	$\delta Q = \pm 0.5°$
Retarder data:	
Transmittance ratio:	$0.95 \leq T_c \leq 1.05$
Phase shift:	$80° \leq \Delta_c \leq 100°$
Angle of incidence error:	$\delta\alpha = \pm 0.1°$

The simulated data, corresponding to readings obtained from an accurate measurement on an imperfect null ellipsometer, were subsequently evaluated under the assumption of an ideal instrument with the parameters of Table 1.

In general, both independent pairs of polarizer/analyzer settings were used in the quantitative determination of the various error effects whose results are shown in Fig. 13. Similar to the evaluation of ellipsometric data in real life, correction of all errors was attempted where this was possible; in particular, the aforementioned approaches for a compensation of errors of the angle of incidence and of tilt errors were used. It should be noted that no tilt error had been assumed in the generation of the simulated "measured" data; however, some of the error effects investigated resulted in a non–zero tilt error when the "measured" data were submitted to the tilt compensation algorithm. This indicates that great care is required in the interpretation of error compensation results which may indicate an error where there is, in fact, none. Generally, the compensation for the tilt

angle β was found to have only a minor influence on the results obtained, except if the error was caused by a phase shift deviation of
the retarder.

(a)

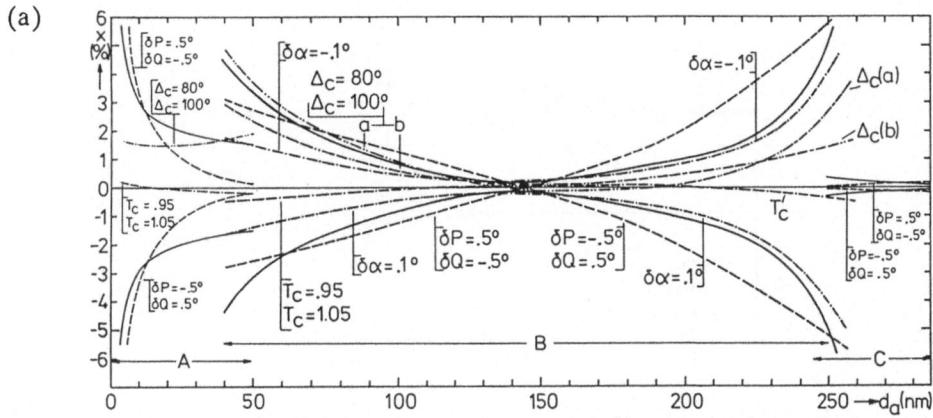

(b)

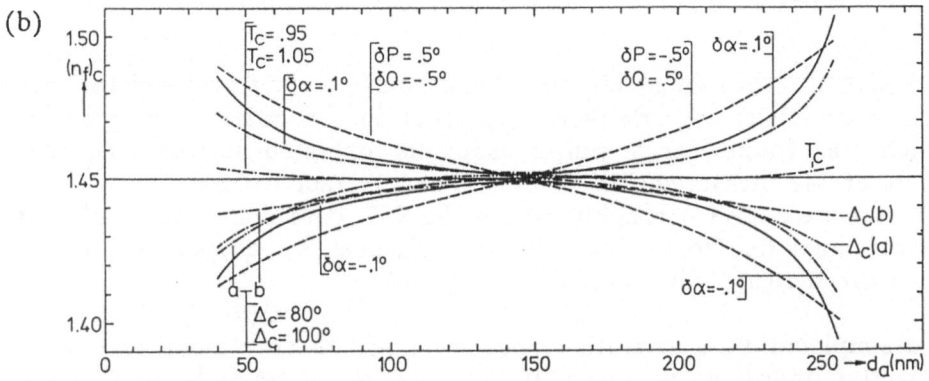

Fig. 13: Alignment errors: Influence of apparatus alignment and component errors on the relative error x of the calculated film
thickness (a), on the calculated film refractive index $(n_f)_c$
(b), and on the absolute error of the thickness of ultra-thin
films (c). (Solid lines: 95% confidence limits for uncertainties
$\delta\Psi = 0.2°$ and $\delta\Delta = 0.4°$ – compare Fig. 11. See next page
for Fig. 13 (c).)

(c)

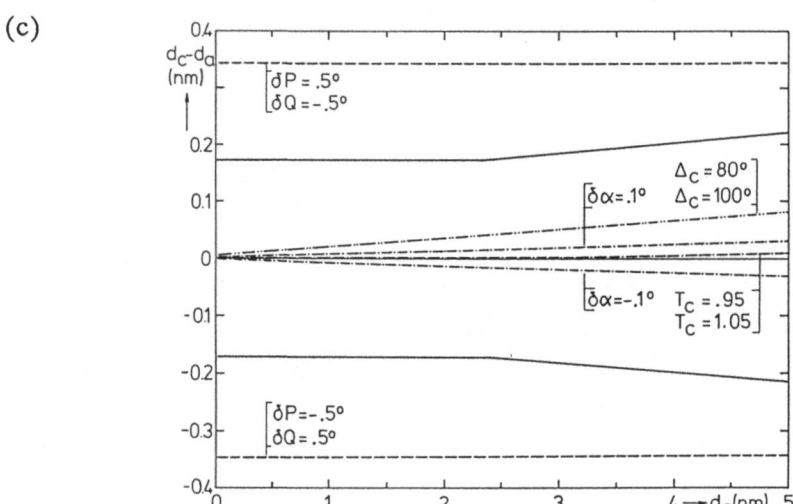

Fig. 13 (c): Alignment errors: Influence of apparatus alignment and
component errors on the absolute error of the thicknesses of
ultra-thin films. (See previous page for a detailed caption.)

Although the instrument alignment and component errors used for the
error estimations are relatively large (but in the order of magnitude
which was found for a simple manually operated instrument), the
errors of the measured film thickness and refractive index lie – with
only one exception – roughly within the 95% confidence limits of the
measurement due to random errors as discussed in chapter 3.1; these
limits are indicated by solid lines in Fig. 13.

The simulated P–A azimuth angles were solved for thickness d and
refractive index n_f in range B, and for thickness only under the
assumption of the correct refractive index n_f in ranges A and C.
With the exception of the curves for an angle of incidence error $\delta\alpha$,
the evaluation routine was used in ranges A and C which varies the
angle of incidence for a minimum imaginary part of the film thickness.
(Doing so for the $\delta\alpha$ curves would have annihilated the effect of this
error altogether.)

A misalignment of the polarizer or the retarder has the most distinct
effect on the calculated film thickness and refractive index data.   In
fact, it is the difference between both azimuths which makes itself
felt most strongly.   A misalignment of the analyzer resulted, in contrast, in such small deviations that they could not even be shown in
Fig. 13; it feigns, however, improbably high tilt angle values.

Table 3: Degree of influence of the various error sources on measured ellipsometric angles, evaluated film parameters, and error terms.

Error	Measured data: Ψ	Δ	Evaluation results: d_c	$(n_f)_c$	d_c^*	Error terms: α	$\mathrm{Re}(\beta)$	$\mathrm{Im}(\beta)$	Notes:
δP	0	3	3–4	3–4	3	1–2	0	0	$P_2=P_1+90°$ $A_2=180°-A_1$
δA	3^a $0–1^b$ $0–1^c$	0^a 0^b $0–1^c$	3^a $0–1^b$ $0–1^c$	3^a $0–1^b$ $0–1^c$	3^a $0–1^b$ $0–1^c$	3^a $0–1^b$ $0–1^c$	3–4	0	$P_2=P_1+90°$ $A_2\neq180°-A_1$
δQ	3^a $1–2^b$ 1^c	3	3^a 3^b 3^c	3^a 3^b 3^c	$3–4^a$ 3^b 3^c	3^a $1–2^b$ $1–2^c$	2–3	0	$P_2=P_1+90°$ $A_2\neq180°-A_1$
δT_c	0^a $1–2^b$	$3–4^a$ 1^b	4^a 2^b	$3–4^a$ $1–2^b$	3^a 1^b	1^a 1^b	1	4	$P_2\neq P_1+90°$ $A_2=180°-A_1$
$\delta\Delta_c$	4^a 3^b 3^c	$2–3^a$ $2–3^b$ 3^c	4^a 3^b 3^c	4^a 3^b 3^c	$2–3^a$ $1–2^b$ 1^c	3^a 1^b 1^c	4–5	0	$P_2=P_1+90°$ $A_2\neq180°-A_1$
$\delta\alpha$	3	3	3	3	0	3	0	0	$P_2=P_1+90°$ $A_2=180°-A_1$

0 ... No influence of error effect.
1 ... Very small influence.
2 ... Small influence.
3 ... Moderate influence.
4 ... Strong influence.
5 ... Very strong influence.

(a) Evaluation based on one pair of P–A values.
(b) Evaluation based on two pairs of P–A values.
(c) Compensation of calculated angle of tilt β prior to evaluation.

* Film thickness calculated with a fixed refractive index n_f.

While an error of the retarder transmittance T_c is only of minor effect, this does not apply to an error of the phase shift Δ_c if no tilt angle compensation is performed (curves $\Delta_c(a)$ in Fig. 13). A significant overall improvement could be achieved by introducing tilt angle compensation (curves $\Delta_c(b)$ in Fig. 13). Tilt angle compensation had almost no effect in range A; in range C, the retarder characteristics were of no consequence whatsoever.

Table 3 gives a qualitative comparison of the influences of the various error effects as obtained from the computer simulations for different evaluation approaches, i.e., for one and two pairs of P–A settings without tilt compensation, and for two pairs of P–A angles with tilt compensation.

3.2.1.2 Numerical Alignment

Besides the wavelength – which is usually well known and constant if sources like lasers or gas discharge lamps are used, and which was therefore not taken into account in the entire error analysis – there are three groups of apparatus parameters which may contribute to a systematic error of an ellipsometric measurement, as reviewed in the previous chapter:

(1) The characteristics of the retarder, i.e., ρ_c (or T_c and Δ_c).

(2) The azimuthal alignment of the polarizer, retarder, and analyzer with respect to one another, and to the plane of incidence.

(3) The angle of incidence of the probe beam.

All these parameters depend to a certain degree on the manufacturing precision of the instrument. Since they explicitly enter into the expressions used for the evaluation of ellipsometric measurements, their deviations from their ideal values can, on principle, be compensated for if they are known. Retarder data can be calculated from two sets of extinction angles for the same sample [7], provided the azimuthal alignment of the polarizer, retarder, and analyzer is correct. It is possible to physically calibrate these azimuths if the instrument can be set to a "straight through" mode where the polarizer, retarder, and analyzer are aligned in one optical axis, omitting the reflection at the sample [62]. However, many simple instruments were designed with a fixed angle between the polarizer and analyzer arms, and do therefore not allow this alignment technique. Furthermore, there is no straightforward way to check the angle between the polarizer and the analyzer arms of such ellipsometers, and therefore the angle of incidence. An alignment of the angle of incidence, if it is feasible at all, requires the use of calibration tools like glass prisms.

Because it is difficult or even impossible to physically align simple null ellipsometers perfectly, a procedure was devised for the calibration of these instruments which relies on numerical correction of the measured data according to the imperfections of the instruments, rather than on mechanical adjustments [14]. A numerical correction can be implemented very easily in a computer program for the measurement evaluation; it can also be applied afterwards to data obtained before the adjustment parameters were determined.

Polarizer, Analyzer, and Retarder Azimuth Alignment

The first step in the numerical alignment procedure comprises the determination of possible azimuth errors δP, δA, and δQ of the polarizer, the analyzer, and the retarder, respectively. For this stage, a reasonably large number of reliable measurements of as many different samples as possible is required each of which must comprise both complementary sets of extinction angles, (P_1-A_1) and (P_2-A_2). The character of the samples does not matter at all; the ellipsometric angles Ψ and Δ derived from them by the simple linear combination approach should be spread over the entire $\Psi-\Delta$ plane, if possible. In order to prevent wedge errors from affecting the alignment procedure, the samples should be as flat and parallel as feasible.

On principle, the parameters of the retarder, ρ_c or T_c and Δ_c, can be calculated from the two sets of polarizer and analyzer azimuths for extinction according to [7]:

$$T_c = [-\tan (P_1 - Q) \cdot \tan (P_2 - Q)]^{-\frac{1}{2}} \tag{8}$$

$$\tan \Delta_c = \frac{[-B^2 - \tan (P_1 - Q) \cdot \tan (P_2 - Q)]^{\frac{1}{2}}}{B} \tag{9}$$

with

$$B = \frac{\tan A_1 \cdot [\tan (P_2 - Q) \cdot \tan Q - \tan (P_1 - Q)/\tan Q]}{2 \cdot (\tan A_1 - \tan A_2)} -$$

$$- \frac{\tan A_2 \cdot [\tan (P_1 - Q) \cdot \tan Q - \tan (P_2 - Q)/\tan Q]}{2 \cdot (\tan A_1 - \tan A_2)} \tag{10}$$

The above equations have to be solved for each measured set of azimuth values. If P_1, A_1, P_2, and A_2 were the correct azimuth angles for extinction, and if Q were the correct azimuth setting of the retarder, exactly the same retarder parameters T_c and Δ_c would result for all sets. Practically, this will not be the case due to alignment errors; major fluctuations of the calculated retarder data will usually ensue in this case. If the measured extinction angles are sorted, for example, in ascending Δ order, the retarder parameters derived from them will follow a certain trend, superimposed by random fluctuations which are due to the inevitable statistical errors of the measurements.

If the above procedure is repeated, with small correction values δP, δA, and δQ being added to P_i, A_i, and Q, respectively, the trends of T_c and Δ_c will change in general; they can be minimized by choosing the proper correction angles. This can be done by means of a computer routine for a three–dimensional search for a minimum, or simply by inspection of the resulting data and according "manual" choice of the correction values. Finally, the calculated T_c and Δ_c data will only exhibit small random fluctuations; their mean values can be regarded as the correct retarder parameters.

The correct azimuth angles for any arbitrary measurement can be computed now as the sums of the scale readings and the pertinent correction angles obtained for minimum retarder parameter fluctuations. The exact transcendental relation between the azimuths and the complex reflection coefficient (eq. (6)), and T_c and Δ_c derived from the above calibration procedure, should be used for calculating the ellipsometric angles Ψ and Δ.

The determination of the retarder parameters and the alignment angles does not depend at all on the angle of incidence of the probe beam. In fact, the measurements used in the alignment procedure may even have been taken at various angles of incidence if the instrument allows to do so.

Angle of Incidence Alignment

Deviations of the actual angle of incidence from the value used for
the evaluation of ellipsometric measurements on film–covered samples
do not matter at all if the refractive index of the film is known, and
if the measurements are analyzed with the algorithm presented in the
preceding chapters where the computational angle of incidence value is
varied until the imaginary part of the calculated thickness disappears.
This is obviously not possible for those measurements where the
thickness <u>and</u> the refractive index of a film are to be determined; an
uncompensated error of the angle of incidence constitutes an important
error mechanism in this case (compare Fig. 13). Still, a calibration of
the angle of incidence can easily be carried out using an ellipsometric
measurement of a sample covered with a single homogeneous dielectric
film, preferably of known composition. For the sake of the accuracy
of this calibration step, the proper retarder and azimuth alignment
data should already have been resolved.

There would be no restriction to the thickness of the dielectric film
used in the angle of incidence calibration procedure if its refractive
index were known exactly. The imaginary part of the calculated film
thickness passes through zero at the correct angle of incidence for
arbitrary film thicknesses, similar to the special case of an ultra–thin
film shown in Fig. 14, if the correct film refractive index is used in
the evaluation procedure. Since the actual refractive index of a film
depends strongly on the technological parameters of its production,
this prerequisite can hardly be fulfilled practically; it might be pos–
sible to calibrate the test film with another ellipsometer or with an
entirely different experimental technique, but such an approach will
most likely introduce new errors. While, therefore, the refractive
index of a thin film is accurately known in very rare cases only, this
does not apply to substrate parameters. At least for semiconductor
grade materials, the published parameters are reasonably reliable;
particularly for single crystal silicon substrates with a low doping
density and for wavelengths in the visible range, the actual doping
density and the type of impurities is unlikely to affect an ellipsometric
measurement. (Only doping densities considerably greater than about
10^{19} cm^{-3} were published to be detectable ellipsometrically [5],[26].)
Hence, any approach which is based upon substrate rather than film
parameters is preferable for a calibration procedure. It appears plau–
sible that the refractive index of the substrate will increasingly domi–
nate the behavior of a system if the thickness of a film approaches
zero.

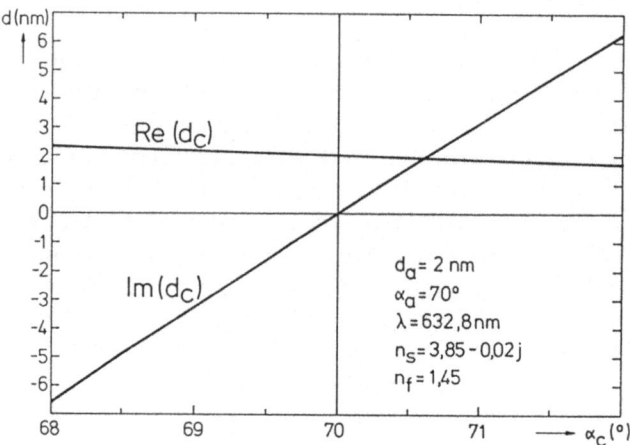

Fig. 14: Angle of incidence determination: Real and imaginary parts
          of the calculated film thickness d_c as functions of the angle
          of incidence α_c used in the calculations.

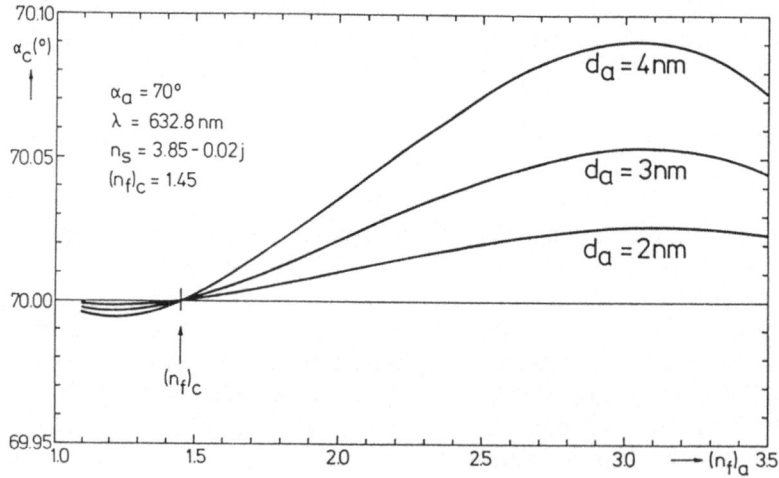

Fig. 15: Angle of incidence determination: Calculated angle of inci-
          dence α_c as a function of the actual film refractive index
          $(n_f)_a$.

A computer simulation was made in order to assess the influence of a
possible uncertainty of the refractive index of thin films used for
angle of incidence calibration. Ψ–Δ data were calculated for various

assumed thicknesses d_a and refractive indices $(n_f)_a$ of a film on a silicon substrate, and submitted to the evaluation procedure for the angle of incidence outlined above. During the latter step, the refractive index of the film $(n_f)_c$ was set to 1.45, corresponding to pure silicon dioxide; the same silicon substrate parameters were used as for the initial calculation of Ψ and Δ. The results of these simulations are shown in Fig. 15; they prove that a reasonable accuracy of the angle of incidence determination can only be achieved with ultra–thin films. The error caused by a deviation of the actual refractive index of the film from the value used for its analysis increases steeply for greater film thicknesses. In contrast, even large refractive index deviations hardly affect the angle of incidence calibration if the films are thin enough. Incidentally, moderate absorption of the films tends to improve the results.

Ultra–thin dielectric films on silicon substrates can easily be prepared: After treatment of a silicon wafer with an etch which removes oxide layers (e.g., with hydrofluoric acid), a native oxide film is formed immediately when the samples are exposed to air at room temperature. The thickness of this native oxide film ranges between 2 and 3 nanometers; it remains almost stable over a prolonged period of time after its initial fast growth. The stability of the films can be enhanced if the samples are immersed in boiling deionized water after etching [63]. The composition and structure of such films, and therefore their refractive indices, depend probably strongly on the actual sample treatment; there is evidence that native oxide films consist of silicon monoxide (SiO) rather than SiO_2 [64]. According to the results shown in Fig. 15, the angle of incidence still can be determined with an accuracy of a few hundredths of a degree if an index of 1.45 is used for its computation. Practically, the refractive index of an unknown native oxide film may lie anywhere between the refractive indices of vacuum (1.00) and of the substrate (3.85 – 0.02j), obviously excluding both boundary values: a porous silicon dioxide film will have a refractive index less than 1.45, while the index of a film which contains excess silicon will range between 1.45 and 3.85. The exact index of a composite film can be calculated from the Law of Clausius and Mosotti [65],[66]; this issue will be discussed in more detail in chapter 3.3.

In order to estimate the uncertainty of a particular calibration procedure, the actual film thickness ought to be known. The calculated thickness value depends, however, again on the actual refractive index of the film. The dependence of the calculated thickness on the actual index is outlined in Fig. 16. Although the measured film thickness may differ significantly from the actual film parameter, it permits still an estimation of the reliability of the angle of incidence calibration:

According to Fig. 16, the strongest deviations occur for actual film refractive indices less than 1.45; in this regime, the actual film thickness may be significantly greater than the thickness value measured ellipsometrically. The impact of the film thickness on the angle of incidence calculation is very small in this range, though (compare Fig. 15). In contrast, the actual film thickness is less than the measured value if the actual refractive index is within the range $1.45 \leq (n_f)_a \leq 2.7$; using Fig. 15 for determining the angle of incidence uncertainty will therefore give a safe estimation for films in this refractive index range. Film refractive indices greater than 2.7 are not very probable any more; such films would have to consist of a mixture of less than 25% SiO_2 and more than 75% pure silicon. For a film with a measured thickness of 2.5 nm (which is about the thickness of the native oxide films formed in air at room temperature) whose refractive index is supposed not to exceed 2.0 (corresponding to a 40% silicon content in a silicon dioxide film), the calculated angle of incidence will lie within $\pm 0.01°$ of its true value.

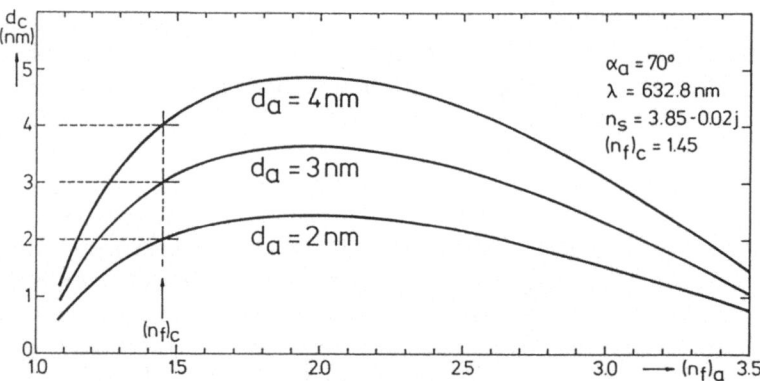

Fig. 16: Angle of incidence determination: Film thickness d_c computed at the calculated angle of incidence α_c as a function of the actual film refractive index $(n_f)_a$.

Since the angle of incidence is also affected by a wedge error of the sample, it is important that the nature and preparation of the calibration standard – for example, a simple silicon wafer – rules out such errors; advantageously, measurements should be taken with various orientations of the wafer on the instrument's stage, and calibration data should be calculated from the mean values derived from these measurements.

The entire numerical calibration procedure can easily be accomplished on a desktop computer; no unusual or expensive calibration standards (except a plain silicon wafer) are required, and no interference with the mechanical alignment of the instrument is necessary. The correction of measured data is likely to significantly reduce systematic errors in measurements done with null ellipsometers, particularly in the critical thickness regions of ultra-thin films and in the vicinity of the boundaries between ellipsometric orders. In many instances, experimental data in these ranges which had been obtained on a simple ellipsometer could not be evaluated at all unless they were corrected according to the results of the numerical calibration.

3.2.2 Rotating Analyzer Ellipsometers

The accuracy of a rotating analyzer ellipsometer is determined by the following elements:

* The mechanical precision and stability of the setup, which shall not be further addressed here. A deviation of the angle of incidence results exactly in the same errors as outlined in the preceding chapters for null ellipsometers; the same technique can be applied for its determination and correction.

* The intrinsic accuracy of the data acquisition and processing system.

* The influence of imperfections of and interferences in the measurement system such as:

  – Orbiting of the angular encoder.

  – Optically or electrically generated noise.

Imperfections of the optical components, and the phase shift induced by the analog photodetector circuitry as discussed in detail by Aspnes et al. [17],[18] are disregarded here on purpose since they can be compensated for numerically.

Because of the transcendental relation between the ellipsometric angles Ψ and Δ and the data primarily measured in rotating analyzer ellipsometry, namely, the shape and azimuth of the polarization ellipse, the above error effects will be surveyed in two parts: First, their influence on the polarization ellipse will be determined quantitatively; subsequently, the effect on Ψ and Δ of errors in the calculated ellipticity and azimuth of the polarization ellipse will be addressed, independent of their origin.

A measurement on a rotating analyzer ellipsometer is based upon the determination of the ellipticity and azimuth of the wave reflected from the sample, which can, in turn, be derived from the intensity modulation depth of the light falling on the photodetector behind the analyzer, and from the phase shift of this intensity modulation with respect to the zero azimuth of the analyzer (compare Fig. 6 and chapter 1.2.2). The light intensity I at the photodetector, and, accordingly, the photocurrent generated by the detector, complies with the following relations:

$$I = I_o \cdot [1 + a\cdot\cos\,(2\omega t) + b\cdot\sin\,(2\omega t)] \qquad (11)$$

or

$$I = I_o \cdot [1 + e\cdot\cos\,(2\omega t + \Phi)] \qquad (12)$$

with

$$e = (a^2 + b^2)^{\frac{1}{2}} \qquad (13)$$

$$\Phi = \text{arctg}\,(-b/a) \qquad (14)$$

where ω is the angular speed of the analyzer prism. The coefficients I_o, a, and b (or I_o, e, and Φ) can be derived from the measured signal by means of a Fourier analysis. This is usually done with digital techniques, in which case the Fourier integrals have to be replaced by the sums of weighted discrete intensity samples. If I_ν is the ν-th data sample $(0 \leq \nu \leq n - 1)$ of n samples per period, the Fourier coefficients can be calculated from:

$$I_o = \sum_\nu I_\nu/n \qquad (15)$$

$$a = \sum_\nu [I_\nu\cdot\cos\,(2\pi\nu/n)]/(\sum_\nu I_\nu/2) \qquad (16)$$

$$b = \sum_\nu [I_\nu\cdot\sin\,(2\pi\nu/n)]/(\sum_\nu I_\nu/2) \qquad (17)$$

3.2.2.1 Intrinsic Accuracy

In order to assess the intrinsic accuracy of the algorithms described above, a digitized intensity signal was simulated with a test program, and submitted to the Fourier transformation routines of a high–speed rotating analyzer ellipsometer [42],[60]. This approach has the advantage of taking into account numerical effects like round–off errors which can hardly be handled by an analytic error estimation. The average intensity I_0 (i.e., the DC component of the signal), the normalized amplitude e (the amplitude of the superimposed AC component divided by I_0), and the phase angle Φ were calculated and compared to the values assumed in the data simulation. The intensity samples I_ν were simulated according to

$$I_\nu = I_{oo} \cdot [1 + e_o \cdot \cos (2\pi\nu/n + \Phi_o)] \tag{18}$$

where I_{oo}, e_o, and Φ_o are the assumed values for I_0, e, and Φ, respectively. The simulated values were in the range $0 \leq I_\nu \leq 32767$; in order to model the 12–bit resolution of the analog–to–digital (A/D) converter of the particular instrument in mind, they were truncated to the next multiple of 8 less than or equal to I_ν. In general, one parameter (e. g., the phase angle Φ_o) was varied within one set of simulations; the errors

$$\Delta I_o = I_o - I_{oo} \tag{19}$$

$$\Delta e = e - e_o \tag{20}$$

and

$$\Delta\Phi = \Phi - \Phi_o \tag{21}$$

obtained for each of the ten to twenty simulation runs per set were submitted to statistical treatment. This appeared justified since the errors were found to exhibit a quasi–random behavior.

The intrinsic accuracy of the data acquisition system was determined for a wide range of normalized signal amplitudes e_o (0.05 – 1), of average intensities I_{oo} between 64 and 16000, corresponding to resolutions between 4 and 12 bits, and for 32 to 1024 data points per signal period. The general results obtained were:

(a) The "measured" average intensity I_o was always less than its assumed value I_{oo}. This can be attributed to the fact that the conversion of the (floating–point) simulation data to the integer values submitted to the Fourier analysis was done by truncating rather than rounding. The relative error $\Delta I_0/I_{00}$ was almost exactly equal to the resolution of the integer data (e.g., $1/4096 = 0.024\%$ for a simulated 12 bit A/D conversion) (Fig. 17).

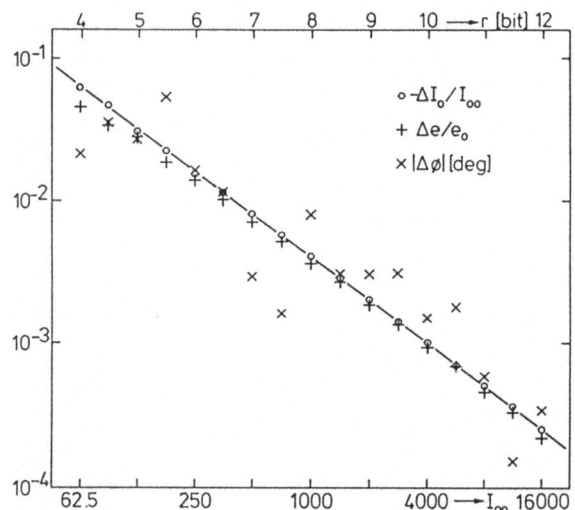

Fig. 17: Intrinsic accuracy of a rotating analyzer ellipsometer: Relative errors of the DC component, ΔI_0, and of the normalized AC amplitude, Δe, and standard deviation of the phase error $\Delta\Phi$ for $e = 1$ as a function of the initial mean intensity I_{00} or of the resolution r of the digitized intensity signal.

(b) Mainly as a consequence of (a), the "measured" normalized amplitude e was greater than its assumed value e_0. Particularly for larger e values ($e > 0.1$), the relative error of e matched perfectly the absolute value of the corresponding relative error of I_0 (Fig. 17). There are deviations from this result for small e values; still, the absolute error of e was found to decrease for smaller normalized amplitudes. Although the error of e depends on the phase angle Φ_0 in a somehow erratic way, its fluctuations lie in the same order of magnitude as its mean value or below; they become more prominent for smaller absolute and normalized signal amplitudes (I_{00} and e_0, respectively), and for a smaller number of data points processed.

(c) The mean error $(\Delta\Phi)_m$ of the phase angle, averaged over the errors obtained for Φ_o in the full range of $0 \leq \Phi_o \leq 2\pi$, was found to be zero under all conditions. However, $\Delta\Phi$ depends on the initial phase angle Φ_o in a similarly irregular way as e; therefore, its standard deviation was regarded as a measure for the phase error. The phase error is approximately inversely proportional to the normalized signal amplitude e and to the number of data points processed; although the correlation between the absolute signal amplitude I_{oo} and $\Delta\Phi$ (Fig. 17) is less pronounced than in the cases of Δe and ΔI_o, a regression analysis discloses a similar behavior. For a resolution of 12 bits ($I_{oo} \approx 16000$), phase errors range between less than 0.001 degrees (e > 0.5, 1024 data points per period) and 0.02 degrees (e = 0.1, 64 data points). Obviously, the angular spacing between two adjacent sampling points does <u>not</u> constitute the limit of the phase angle resolution.

3.2.2.2 Angular Encoder Orbiting

An error mechanism which cannot be avoided entirely in an actual instrument is caused by the mechanical and geometrical imperfections of the angular encoder which defines the analyzer azimuths where intensity samples are taken.

Practically, the encoder bars are never perfectly equidistant. Small random fluctuations are likely to induce an error similar to a noise signal. The amplitude of this noise signal can be estimated as the product of the difference angle between the ideal and the actual positions of the encoder bars, and the maximum derivative of the sinusoidal intensity signal with respect to the analyzer angle. This effect has to be taken into account when the influence of random noise is discussed.

An entirely different error mechanism, however, is caused by an eccentrically mounted and therefore orbiting encoder disk. The apparent interval angle between two encoder bars varies approximately sinusoidally in this case, causing, despite of a constant angular speed ω of the encoder, a frequency modulation of the sampling interval Δt. For a small eccentricity Δr of an encoder with a "working" radius r, the actual modulated sampling interval Δt_m results as

$$\Delta t_m = \Delta t \cdot [1 + \frac{\Delta r}{r} \cdot \cos{(\omega t + \Phi_\delta)}] \qquad (22)$$

where Φ_δ is an arbitrary phase angle which depends on the orientation of the encoder disk with respect to its eccentricity. The time t_ν at which the ν-th sample is taken can be written as

$$t_\nu = \sum_{i=1}^{\nu} (\Delta t_m)_i \tag{23}$$

We can approximate the summation in (23) by an integration; with (22), we can obtain for the difference δt_ν between the actual sampling time t_ν and the proper sampling time $\nu \cdot \Delta t$:

$$\delta t_\nu = \frac{\Delta r}{r \cdot \omega} \cdot \sin (\pi \nu / n + \Phi_\delta) \tag{24}$$

For the ν-th sample, the intensity actually measured will be:

$$I_\nu = I_{oo} \cdot [1 + e_o \cdot \cos (2\omega t_\nu + \Phi_o)] =$$
$$= I_{oo} \cdot \{1 + e_o \cdot \cos [2\pi \nu / n + 2 \cdot (\Delta r / r) \cdot \sin (\pi \nu / n + \Phi_\delta) + \Phi_o]\} \tag{25}$$

The second term in the argument of the cosine function in (25) constitutes obviously a phase modulation with a frequency ω, i.e., with half the signal frequency, and with a modulation depth δ:

$$\delta = 2 \cdot (\Delta r / r) \tag{26}$$

The effect of this phase modulation was investigated accordingly with:

$$I_\nu = I_{oo} \cdot \{1 + e_o \cdot \cos [2\pi \nu / n + \Phi_o + \delta \cdot \cos (\pi \nu / n + \Phi_\delta')]\} \tag{27}$$

(Numerically, the phase modulation term was modelled with a cosine rather than a sine function. The phase shift constant Φ_δ which is probably not known anyway need only be replaced by Φ_δ' which differs from Φ_δ by $\pi/2$ in order to convert the sine into a cosine function.) The errors were determined for a constant assumed mean intensity I_{oo}. Φ_o was set to 0, and Φ_δ' was varied over the entire range from 0 to 2π.

The magnitude of the errors according to these simulations depends strongly on the phase relation between the actual and the modulation signals, i.e., on Φ_δ: amplitude and phase errors as functions of Φ_δ follow approximately sine functions which are out of phase by 90 degrees. Since Φ_δ will probably not be known in any practical application, the mean values and standard deviations of Δe and $\Delta \Phi$ over the range $0 \le \Phi_\delta \le 2\pi$ were plotted in Figs. 18 and 19, respectively. (The mean value of $\Delta\Phi$ is zero again.) In addition, a second harmonic with a normalized amplitude e_2 was found to be generated. The mean value and the standard deviation of e_2 are shown in Fig. 20. The magnitude of the errors depends distinctly on the number of signal periods processed. The curves in Figs. 18 to 20 labeled "a" were obtained from a Fourier integration over one signal period (1024 data points), whereas the "b" curves result from an integration over two signal periods (or one full analyzer rotation period, i.e., 2048 data points).

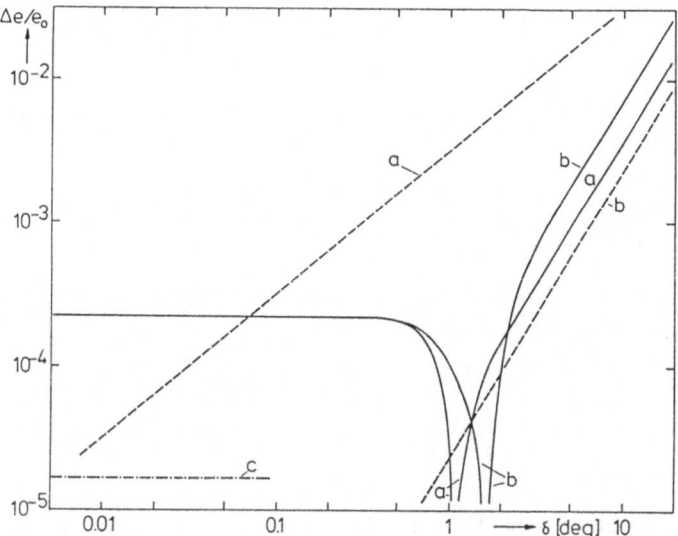

Fig. 18: Effect of angular encoder orbiting: Error Δe of the normalized AC amplitude as a function of the phase modulation depth δ for one (a) and two (b) signal periods processed. (Solid lines: error mean values; broken lines: standard deviation of Δe; line c: standard deviation of Δe for $\delta = 0$. The mean values of the amplitude error are positive for $\delta < (1 - 3)°$, and negative for greater δ values.)

The effect of the phase modulation amplitude on the normalized amplitude error Δe is presented in Fig. 18. The mean values of Δe (solid lines) differ hardly for the cases "a" and "b"; the errors change their signs in the vicinity of $\delta = 1°$. In case "a", however, even small phase modulation depths cause large fluctuations of the error whose standard deviations are shown with broken lines in Fig. 18. These error fluctuations exceed the mean error by far; a phase modulation of less than 0.1° is likely to double the amplitude error. The standard deviation of Δe is proportional to δ. The evaluation of two rather than one signal periods (case "b") comprises a full phase modulation period; it keeps the amplitude error in the order of its "ideal" magnitude up to a phase jitter of more than 1 degree. (For larger δ, the mean value and the standard deviation of Δe follow a δ^2 function if the integration comprises two signal periods.)

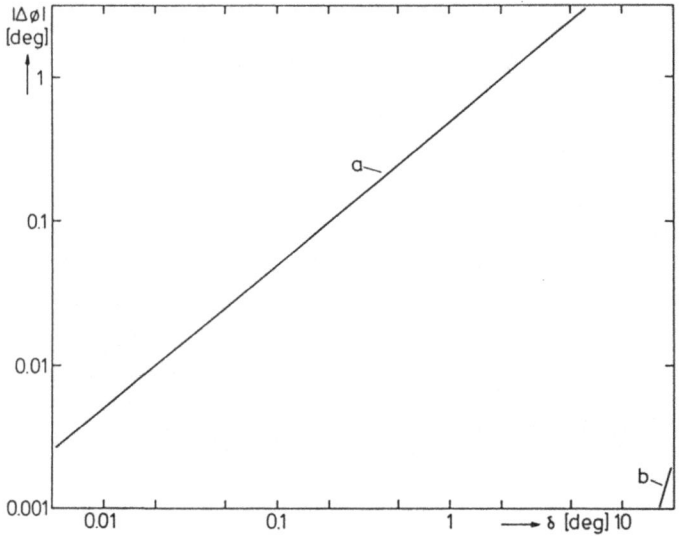

Fig. 19: Effect of angular encoder orbiting: Standard deviation of the phase error $\Delta\Phi$ as a function of the phase modulation depth δ for one (a) and two (b) signal periods processed. (The mean phase error is zero.)

A similar behavior was found for the phase error (Fig. 19) and for the amplitude of the second harmonic (Fig. 20). In both cases, the standard deviation of the phase error $\Delta\Phi$, and the second harmonic amplitude e_2, respectively, are reduced by several orders of magnitude if two signal periods are processed ("b"), compared to their single-period values ("a"). (The "a" curves are proportional to δ, the "b" curves, to δ^2.)

Equation (26) permits to estimate the influence of encoder orbiting on a given system. The modulation depth δ (in radians) is equal to twice the relative eccentricity of the encoder disk. For example, a δ value of close to 0.1° ensues for an encoder with r = 25 mm and an eccen-

tricity of 0.02 mm, which is about standard. The errors caused by encoder orbiting are evidently by several orders of magnitude smaller in
this case if the integration is done over two signal periods rather than
one (compare Figs. 18 – 20).

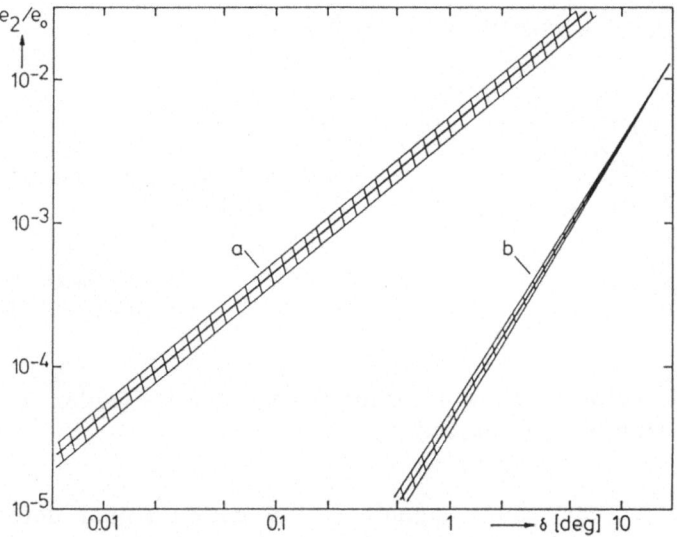

Fig. 20:   Effect of angular encoder orbiting: Normalized amplitude of
the second harmonic e_2 as a function of the phase modulation depth δ for one (a) and two (b) signal periods processed.
(Solid lines: mean values; shaded areas: confidence ranges.)

3.2.2.3 Periodic and Random Noise

The influence of noise was studied for random and for sinusoidal periodic noise. Due to the linearity of the Fourier transformation, the
noise and the actual intensity signals can be regarded in either case
as being separately submitted to the Fourier transformation; the measured combined signal can be obtained from a vector addition of the
harmonic signals derived from the actual intensity and noise signals,
respectively.

We assume the same dependence of the actual intensity signal I_s as
above:

$$I_s = I_{oo} \cdot [1 + e_o \cdot \cos (2\pi\nu/n + \Phi_o)] \tag{28}$$

In the absence of an actual intensity signal, the noise signal alone
would be evaluated by the Fourier algorithm as if it were a sinusoidal
signal with the frequency 2ω and with DC and AC components I_{on} and
$I_{on} \cdot e_n'$, respectively:

$$I_n = I_{on} \cdot [1 + e_n' \cdot \cos (2\pi\nu/n + \Phi_n)] =$$

$$= I_{on} + I_{oo} \cdot e_n \cdot \cos (2\pi\nu/n + \Phi_n) \tag{29}$$

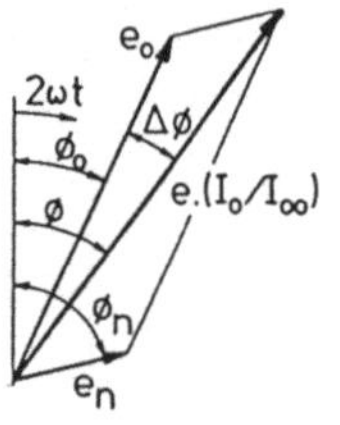

$$I = I_{oo}[1 + e_o \cdot \cos(2\omega t + \phi_o)] + I_{on} + I_{oo} e_n \cos(2\omega t + \phi_n)$$
$$= I_o [1 + e \cdot \cos (2\omega t + \phi)]$$

$$\Delta I = I_o - I_{oo} = I_{on} \qquad |\Delta I / I_{oo}| \ll 1$$
$$\Delta e = e - e_o \qquad (\Delta e / e_o)_{max} \doteq -\Delta I / I_o \pm e_n / e_o$$
$$\Delta \phi = \phi - \phi_o \qquad (\Delta \phi)_{max} = \pm \arctan (e_n / e_o)$$

Fig. 21:   Vector addition of probe beam intensity (index "o") and noise
(index "n") induced signals.

The sum of the above signals is the actual signal detected.   The
resulting errors of e and Φ can be determined from a vector addition
of the two cosine components in eqs. (28) and (29) as shown in Fig.
21.   For this purpose, the AC component of the noise induced signal
must be normalized with the same DC component I_{oo} as the actual
intensity signal, as defined in the second line of eq. (29).   A worst
case analysis is advantageous since the phase relation between the
actual intensity and noise signals is hardly known.   The effects of
periodic and random noise were therefore investigated by calculating
the (normalized) average intensity I_{on} and the total amplitude $I_{on} \cdot e_n$
rather than the actual errors Δe and $\Delta \Phi$.

Periodic Noise

Periodic noise in a rotating analyzer ellipsometer can be caused either
purely electrically (as the result of emf interference), or optically,
e.g., by power line noise superimposed on the laser's power supply, or
by stray light from artificial sources.

Periodic noise was modelled according to

$$I_\nu = I_p/2 \cdot [1 + \cos (2x\pi\nu/n + \Phi_p)] \tag{30}$$

with I_p, the amplitude of the noise signal, and x, the ratio between the signal and noise frequencies.

(a)

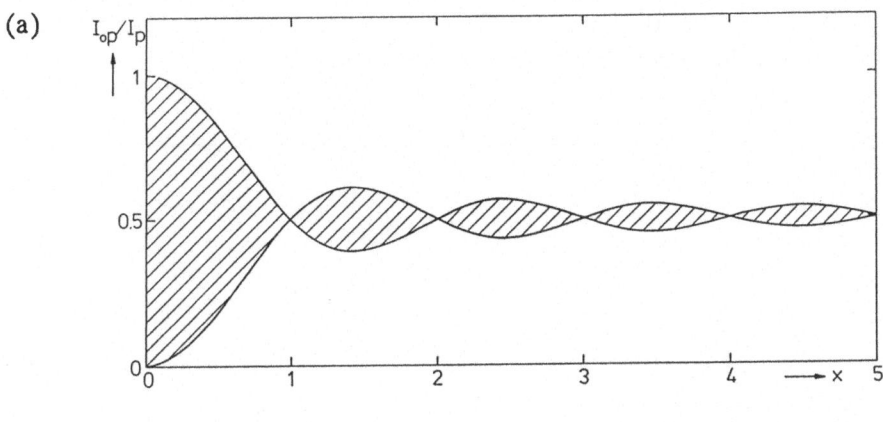

(b)

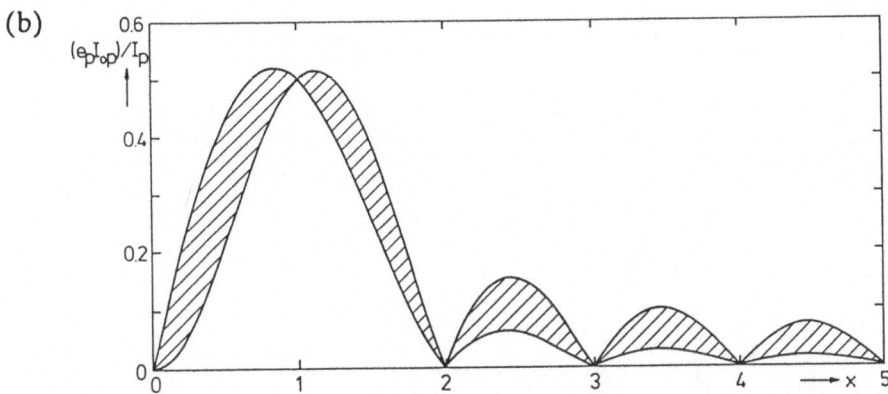

Fig. 22: Effect of periodic noise: DC (a) and AC (b) amplitudes of the detected component with the frequency 2ω for a noise frequency $2x\omega$ and for an integration over one signal period. (The shaded areas indicate the effect of the phase angle Φ_n of the noise oscillation.)

Eq. (30) was integrated analytically. Figs. 22 and 23 show the noise induced DC component I_{op} and the total AC amplitude $I_{op} \cdot e_p$ for an integration over one and two actual signal periods, respectively. The shaded areas denote the range caused by a variation of Φ_p. The results indicate that an integration over several signal periods reduces the bandwidth of the data acquisition system and therefore suppresses periodic noise with a frequency different from the one of the actual intensity signal very efficiently. Noise with an integer number of noise periods within the integration period is totally locked out. The fluctuations of the DC component become smaller, too, if several signal periods are processed, particularly if the noise frequency is greater than the actual signal frequency (i.e., if x > 1); a constant DC component can more easily be compensated for.

(a)

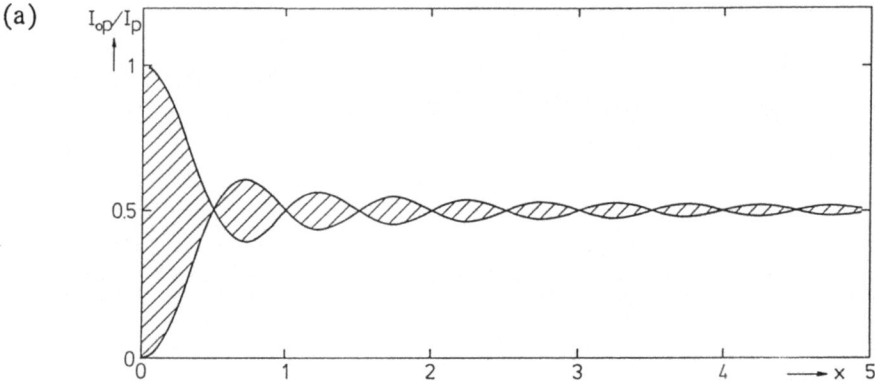

(b)

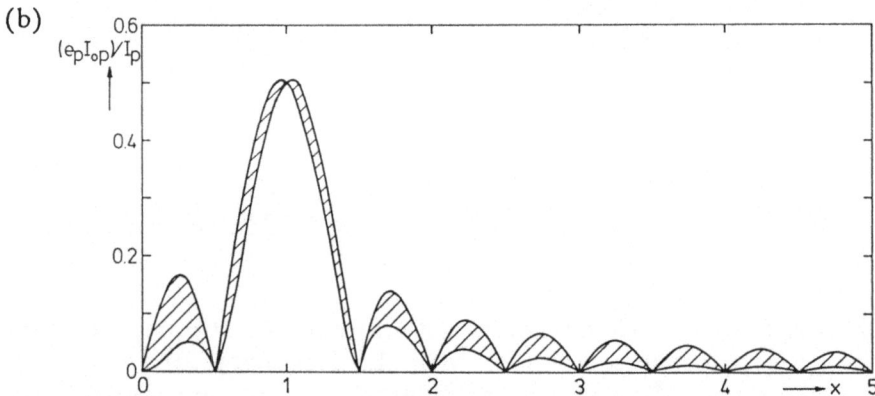

Fig. 23: Effect of periodic noise: DC (a) and AC (b) amplitudes of the detected component with the frequency 2ω for a noise frequency $2x\omega$ and for an integration over two signal periods. (The shaded areas indicate the effect of the phase angle Φ_n of the noise oscillation.)

Random Noise

Random noise was simulated with a numeric random number generator [67] specially adapted for this purpose. It was modelled as

$$I_\nu = I_{oo} + (I_r/2) \cdot r_\nu \qquad (31)$$

with the (peak–to–peak) noise amplitude I_r, and r_ν, a random function $(-1 \le r_\nu \le 1)$.

For small noise amplitudes, an effect of the signal mean value on the measured DC and signal frequency amplitudes I_{om} and e_m, respectively, can be expected due to the signal quantization in the A/D conversion process (Fig. 24): A small fluctuation of the input signal with an amplitude less than the resolution of the A/D converter is detected and results in a non–zero normalized amplitude e_r if the fluctuations happen to pass across a quantization boundary (Fig. 24(a)), while it remains undetected ($e_r = 0$) if the signal lies entirely within one quantization step (Fig. 24(b)). Accordingly, the measured DC component I_{om} does no more follow minor shifts of the original signal in case (b), although it does so in case (a). Small noise amplitudes may therefore pass undetected in the simulations while they may influence the measurements under practical conditions. The bends in the graphs of Figs. 25 and 26 can be attributed to these signal quantization effects.

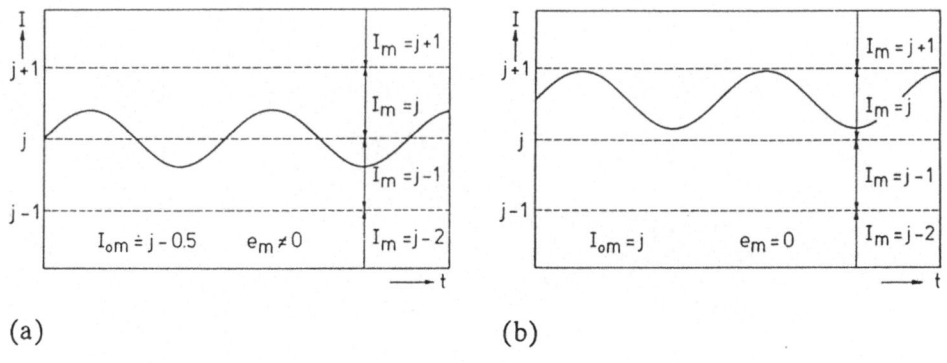

(a)   (b)

Fig. 24: Quantization error for small signals.

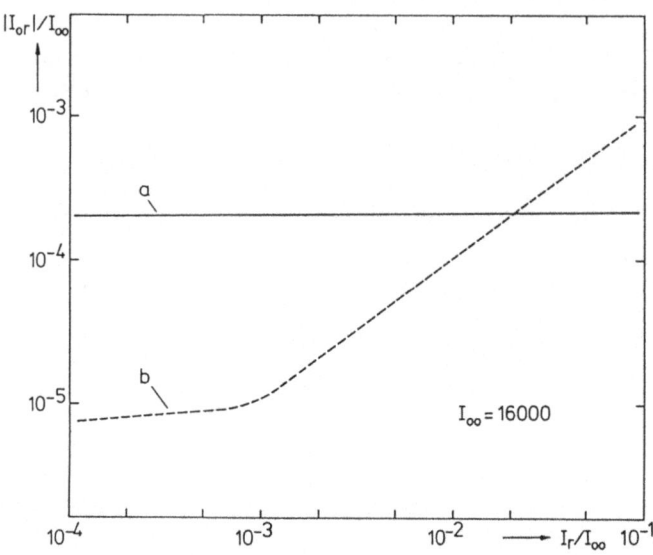

Fig. 25: Effect of random noise: Mean value (a) and standard deviation (b) of the error of the DC component for n = 1024 data
sampling points. (The mean value is always negative.)

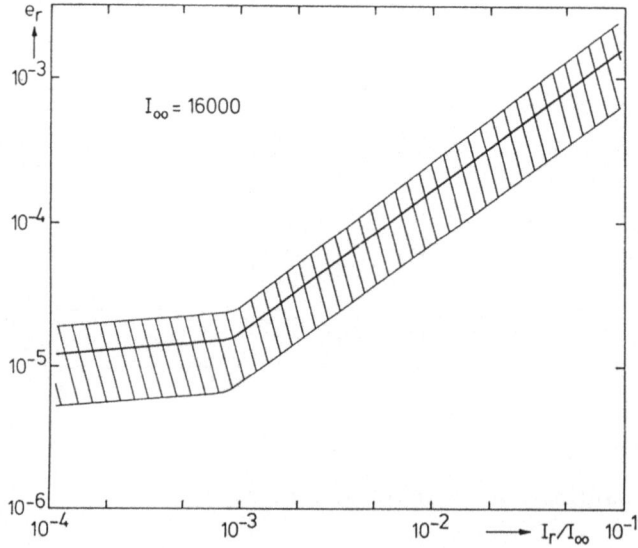

Fig. 26: Effect of random noise on the normalized amplitude e_r of the
detected periodic component for n = 1024 data sampling
points. (The shaded area indicates the confidence range.)

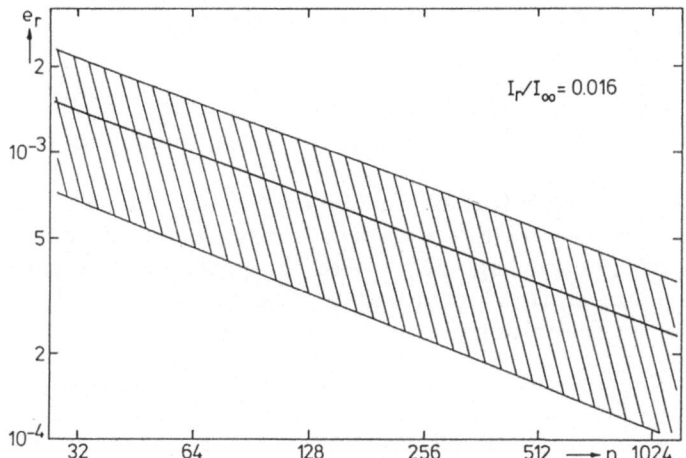

Fig. 27: Effect of random noise on the periodic component with the frequency 2ω as a function of the number n of data points per period processed. (The shaded area indicates the confidence range.)

Figs. 25 and 26 show the influence of the noise amplitude I_r on the DC component I_{or} and on the normalized amplitude, e_r, respectively, which result from a Fourier analysis of the noise signal alone. While the impact on the DC component I_{or} is moderate, a pronounced effect on e_r can be observed. (The standard deviation of I_{or} – curve "b" in Fig. 25 –, and e_r are proportional to the noise amplitude.) The normalized amplitude e_r is inversely proportional to the square root of the number of data points processed, which was to be expected from theory (Fig. 27). A similar behavior could also be observed for the standard deviation of I_{or}.

3.2.2.4 The Influence of the Error Effects Discussed on the Ellipsometric Angles Ψ and Δ

The principal setup of a rotating analyzer ellipsometer as shown in Fig. 6 can be expanded by the introduction of a retarder (i.e., of a quarter-wave plate) immediately after the polarizer, resulting in the same configuration of the polarizer arm as for the null ellipsometer of Fig. 5. In this case, the probe beam incident on the sample is, in general, elliptically rather than linearly polarized. On such a generalized rotating analyzer ellipsometer, the following relation describes the dependence of the ellipsometric angles Ψ and Δ on the measured relative amplitudes a and b of the cosine and sine components of the intensity signal (compare eq. (11) on page 44) [1]:

$$\tan \Psi \cdot \exp(j \cdot \Delta) = \frac{1 + a}{b \pm j \cdot (1 - a^2 - b^2)^{\frac{1}{2}}} \times \frac{\tan Q + \rho_c \cdot \tan(P - Q)}{1 - \rho_c \cdot \tan Q \cdot \tan(P - Q)} \qquad (32)$$

P and Q are the azimuths of the polarizer and the retarder, respectively, and ρ_c is the relative complex transmittance of the retarder, as defined in eq. (7) (on page 32). With the normalized signal amplitude e and the phase shift Φ according to eqs. (11) – (14), the above expression can be re-written:

$$\tan \Psi \cdot \exp(j \cdot \Delta) = \frac{1 + e \cdot \cos \Phi}{-e \cdot \sin \Phi \pm j \cdot (1 - e^2)^{\frac{1}{2}}} \times \frac{\tan Q + \rho_c \cdot \tan(P - Q)}{1 - \rho_c \cdot \tan Q \cdot \tan(P - Q)} \qquad (33)$$

This general relation can be simplified for a retarder-less instrument with linearly polarized incident beam by setting

$$Q = P \qquad (34)$$

$$\rho_c = T_c \cdot \exp(-j \cdot \Delta_c) = 1$$

In this case, eq. (33) can be reduced to:

$$\tan \Psi \cdot \exp(j\Delta) = \frac{1 + e \cdot \cos \Phi}{-e \cdot \sin \Phi \pm j \cdot (1 - e^2)^{\frac{1}{2}}} \cdot \tan P \qquad (35)$$

The following expressions for Ψ and Δ, respectively, can be derived from eq. (35):

$$\tan \Psi = \tan P \cdot \frac{(1 - e^2 \cdot \cos^2 \Phi)^{\frac{1}{2}}}{1 - e \cdot \cos \Phi} \qquad (36)$$

$$\tan \Delta = \pm \frac{(1 - e^2)^{\frac{1}{2}}}{e \cdot \sin \Phi} \qquad (37)$$

(a)

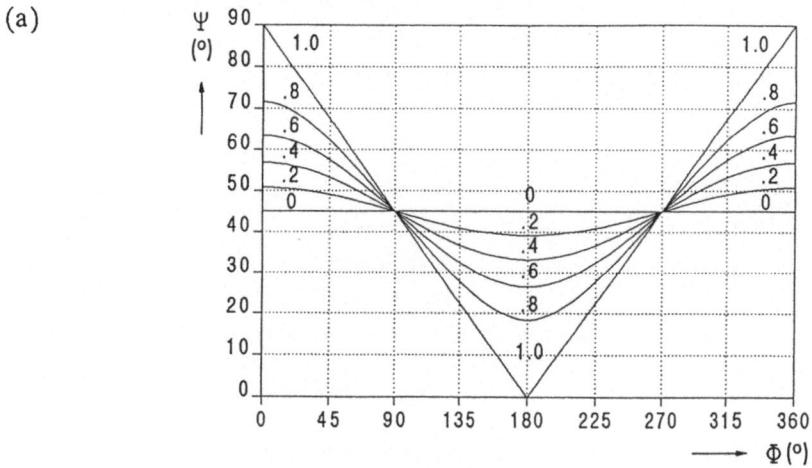

(b)

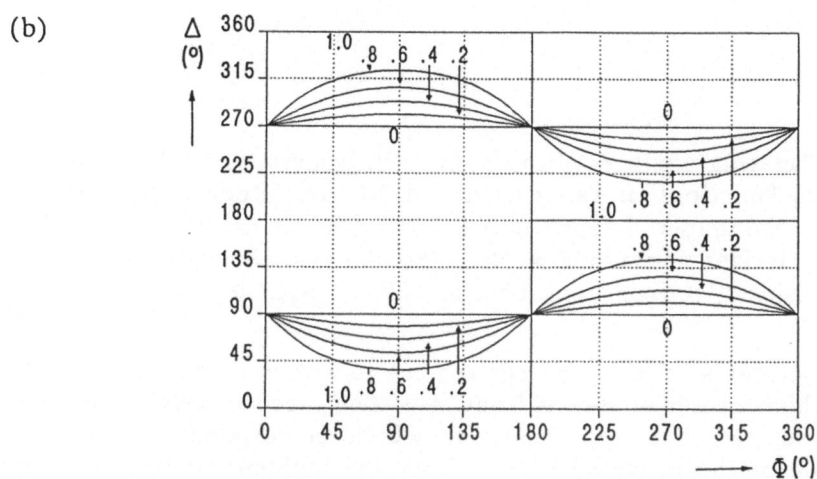

Fig. 28: Rotating analyzer ellipsometry: Ellipsometric angles Ψ (a) and
Δ (b) as functions of the phase angle Φ for linearly polarized
incident light; P = 45°; Parameter: normalized AC amplitude
e.

Obviously, there are two solutions for Δ due to the two possible signs
of the square root in (35). This behavior is to be expected, though,
since the approach chosen in rotating analyzer ellipsometry does not
permit to determine the "handedness" of the polarization ellipse (i.e.,
whether the electric field vector rotates clockwise or counterclockwise). The dependence of Ψ and Δ on e and Φ is shown in Fig. 28 for
linearly polarized incident light and a polarizer azimuth P of 45°

(which is more or less a standard setting); the two diagrams of Fig. 28 have been combined into the Ψ–Δ graph of Fig. 29 where solid lines indicate the Ψ–Δ values for constant normalized AC signal amplitudes e, and broken lines, for constant phase angles Φ.

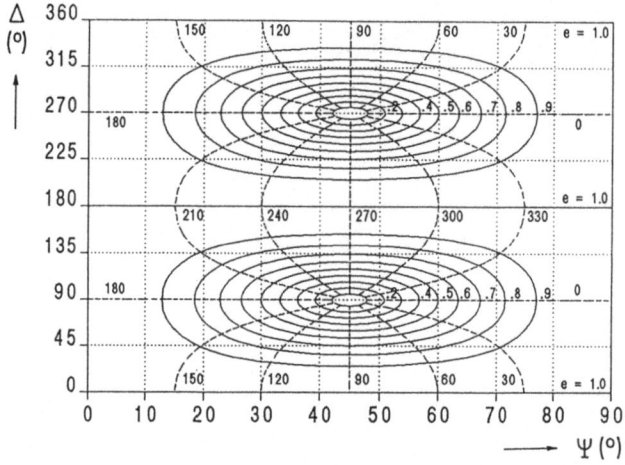

Fig. 29: Rotating analyzer ellipsometry: Ellipsometric angles Ψ and Δ as functions of the normalized AC amplitude e, and of the phase angle Φ for linearly polarized incident light and a polarizer azimuth P = 45°. (Solid lines: constant AC amplitudes e, broken lines: constant phase angle Φ.)

The two solutions of eq. (35) correspond to either of the patterns in Fig. 29 which center at Δ = 90° and Δ = 270°, respectively. Equation (35) degenerates if e approaches 1, which corresponds to a linear polarization of the reflected beam. Since the incident beam is linearly polarized as well, Δ can only be either 0° (or 360°) or 180° in this case. Similarly, eq. (35) degenerates for e = 0, i.e., for a circularly polarized reflected beam. Obviously, the phase angle Φ is irrelevant then, and Ψ is equal to P. (This operation mode is employed in principal angle ellipsometry; compare chapter 1.2.2.)

According to Fig. 29, there are areas in the Ψ–Δ plane where the graphs for constant e and Φ values are relatively sparse. This applies, in particular, to the vicinity of Δ = 0 or 180°, and to small (Ψ < 10°) or large (Ψ > 80°) Ψ values. Small errors of the measured normalized amplitude e and of the phase angle Φ are liable in these regimes to cause major deviations of the calculated ellipsometric angles. This affects particularly the region of ultra–thin films on weakly absorbing substrates (compare Fig. 4). In order to alleviate this problem, the patterns in the Ψ–Δ plane can be shifted closer to the Δ axis by choosing a smaller polarizer azimuth P, as shown in Fig. 30 for P = 15°. Analogously, the accuracy for large Ψ values can be improved by selecting a polarizer azimuth greater than 45°.

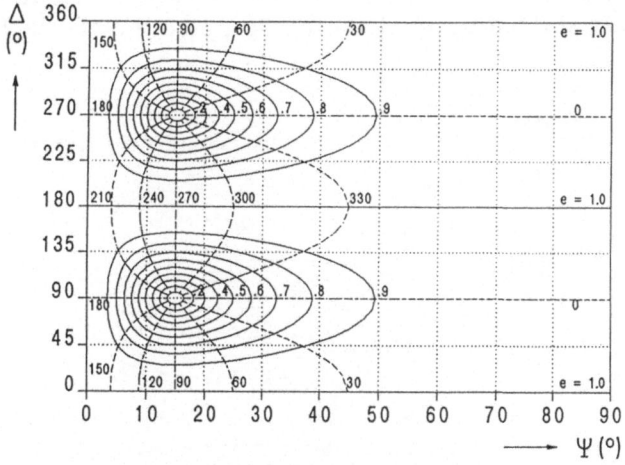

Fig. 30: Rotating analyzer ellipsometry: Ellipsometric angles Ψ and Δ as functions of the normalized AC amplitude e, and of the phase angle Φ for linearly polarized incident light and a polarizer azimuth P = 15°. (Solid lines: constant AC amplitudes e, broken lines: constant phase angle Φ.)

A further improvement can be accomplished by using an elliptically or circularly polarized incident probe beam. This requires, evidently, the use of a retarder immediately after the polarizer, and introduces new problems due to the imperfections of the retarder and its alignment, and due to a wavelength dependence of the retarder parameters. With an ideal quarter–wave plate mounted under 45° with respect to the polarizer azimuth (which results in a circularly polarized incident beam), the following relations ensue:

$$|P - Q| = 45°$$

$$\rho_c = T_c \cdot \exp(-j \cdot \Delta_c) = \pm j \tag{38}$$

(The two signs of ρ_c correspond to the two possible orientations of a quarter–wave plate which produce a circular polarization of the emerging beam.) Substituting the above conditions into eq. (33) results in:

$$\tan \Psi \cdot \exp (j\Delta) = \pm j \cdot \frac{1 + e \cdot \cos \Phi}{-e \cdot \sin \Phi \pm j \cdot (1 - e^2)^{\frac{1}{2}}} \tag{39}$$

With the exception of the factor of $\pm j$, corresponding to a shift of Δ by $\pm 90°$, this result is identical to the one obtained for a linear polarization of the incident beam, and $P = 45°$ (compare eq. (35)). The patterns in the Ψ–Δ plane of Fig. 29 for constant e and Φ are therefore only shifted up or down by $90°$, as indicated in Fig. 31.

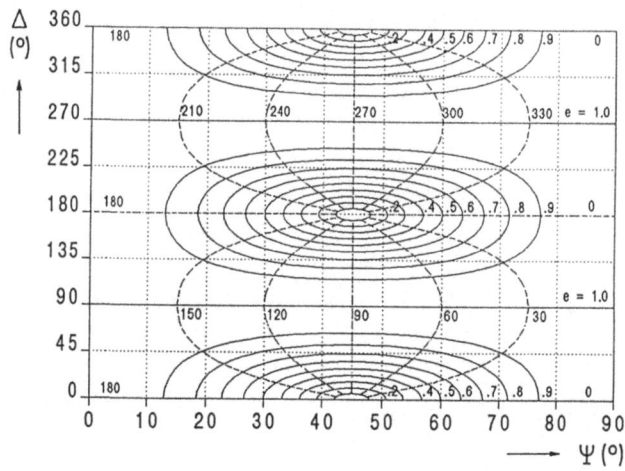

Fig. 31: Rotating analyzer ellipsometry: Ellipsometric angles Ψ and Δ as functions of the normalized AC amplitude e, and of the phase angle Φ for circularly polarized incident light. (Solid lines: constant AC amplitudes e, broken lines: constant phase angle Φ.)

The influence of errors of the measured normalized intensity e and the phase angle Φ was estimated with an approach similar to the one described in the preceding chapters: Either e or Φ or both were varied within a certain range, and the difference between the maximum and minimum resulting Ψ and Δ values was computed. Although the errors Δe and $\Delta \Phi$ used in the simulations were chosen relatively large, the results can be scaled appropriately for smaller errors.

The total uncertainties (i.e., the lengths of the error bars) of Ψ and Δ are shown in Fig. 32 as a function of Φ for various normalized amplitudes e, and for a normalized error $\Delta e/e = \pm 0.01$. The phase error $\Delta \Phi$ was set to zero in this estimation.

(a)

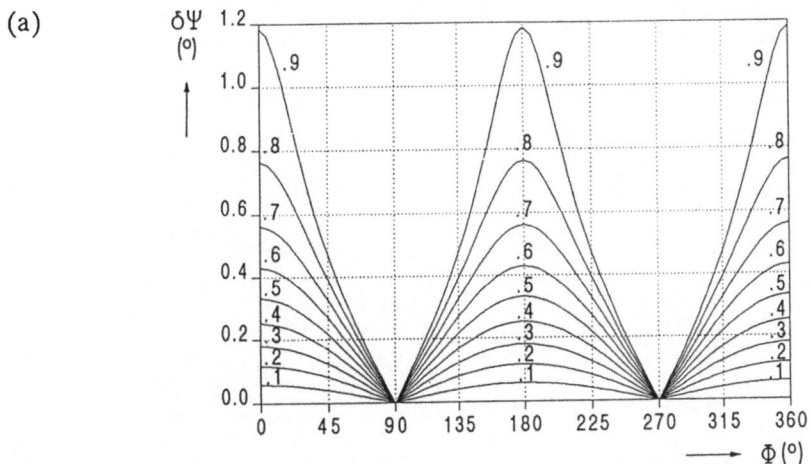

(b)

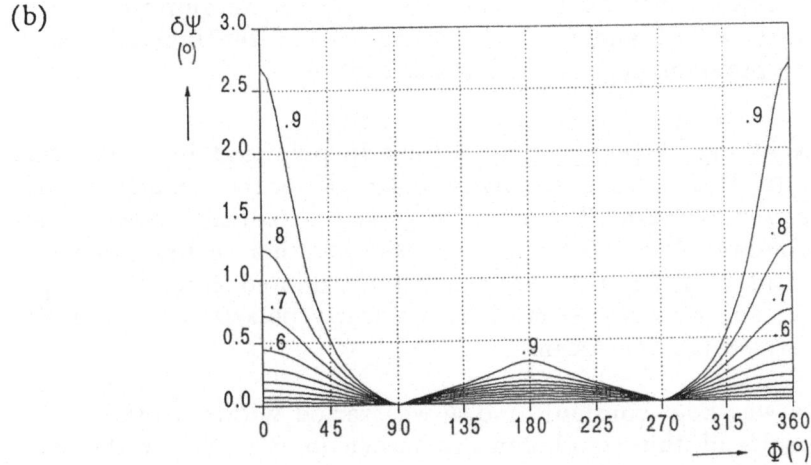

Fig. 32: Rotating analyzer ellipsometry: Uncertainties $\delta\Psi$ (a), (b) and $\delta\Delta$ (c) of the ellipsometric angles for an amplitude error $\Delta e/e$ = 0.01 as a function of the phase angle Φ for linearly polarized incident light with P = 45° (a) and P = 15° (b), or for circularly polarized incident light (a); Parameter: normalized AC amplitude e. The graphs for $\delta\Delta$ (c) apply to any polarization of the incident beam. (The normalized amplitude e was varied in steps of 0.1 between 0.1 and 0.9. See next page for Fig. 32 (c).)

(c)

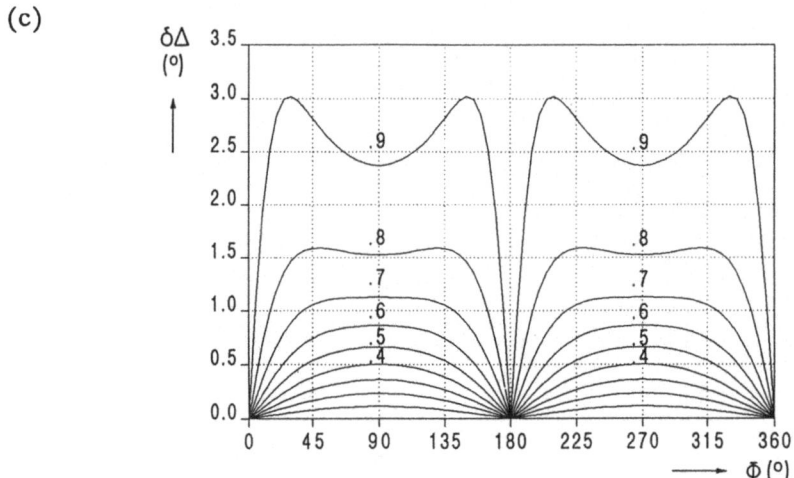

Fig. 32 (c): Uncertainty $\delta\Delta$ as a function of Φ for an amplitude error $\Delta e/e = 0.01$ and for any polarization of the incident beam; see previous page for a detailed caption.

The curves in Fig. 32(a) apply to a linearly polarized incident beam with P set to 45°, or to a circularly polarized probe beam. A different polarizer azimuth (P = 15° in Fig. 32(b)) reduces errors in one area (here, around $\Phi = 180°$) at the expense of other regions. Since Δ is independent of P for the special cases considered here, the graphs in Fig. 32(c) apply to any azimuth of a linearly polarized, or to a circularly polarized incident beam.

Similarly, e was kept constant, and Φ was varied within a range $\Delta\Phi = \pm1°$; the results of this simulation are shown in Fig. 33 for the same cases as above.

Finally, errors of e <u>and</u> Φ were permitted within the ranges of ±0.01 and $\pm1°$, respectively. Their combined impact on Ψ and Δ was determined with an approach similar to the "error ellipse" outlined in chapter 3.1. The resulting "error profiles" of Fig. 34 can roughly be approximated for a particular combination of e and Φ by the maximum of the errors $\delta\Psi$ or $\delta\Delta$ shown in the graphs of Figs. 32 and 33, as indicated for one graph in Fig. 34, which justifies an independent treatment and scaling of either error source.

(a)

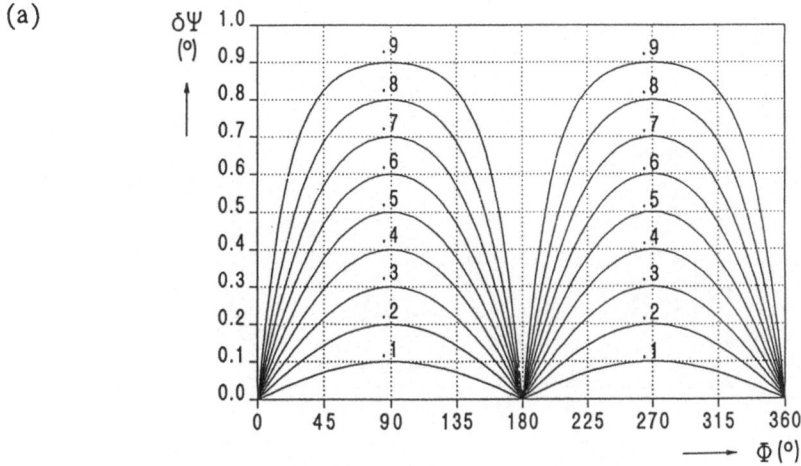

(b)

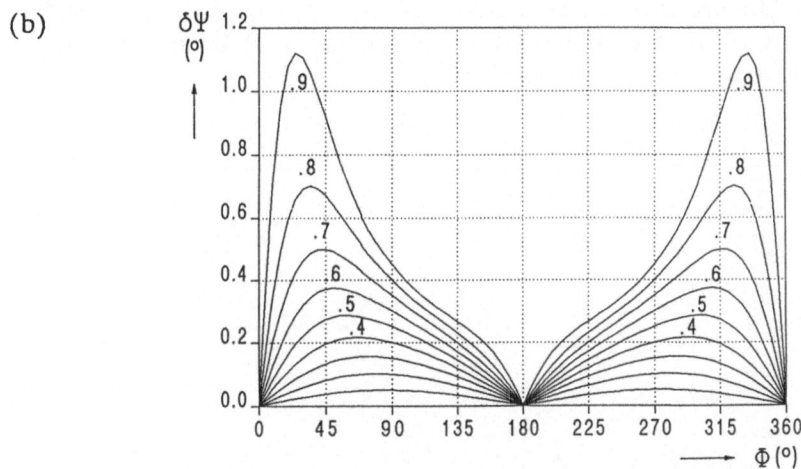

Fig. 33: Rotating analyzer ellipsometry: Uncertainties δΨ (a), (b) and
δΔ (c) of the ellipsometric angles for a phase error ΔΦ = 1°
as a function of the phase angle Φ for linearly polarized
incident light with P = 45° (a) and P = 15° (b), or for circularly polarized incident light (a); Parameter: normalized AC
amplitude e. The graphs for δΔ (c) apply to any polarization
of the incident beam. (The normalized amplitude e was varied in steps of 0.1 between 0.1 and 0.9. See next page for
Fig. 33 (c).)

(c)

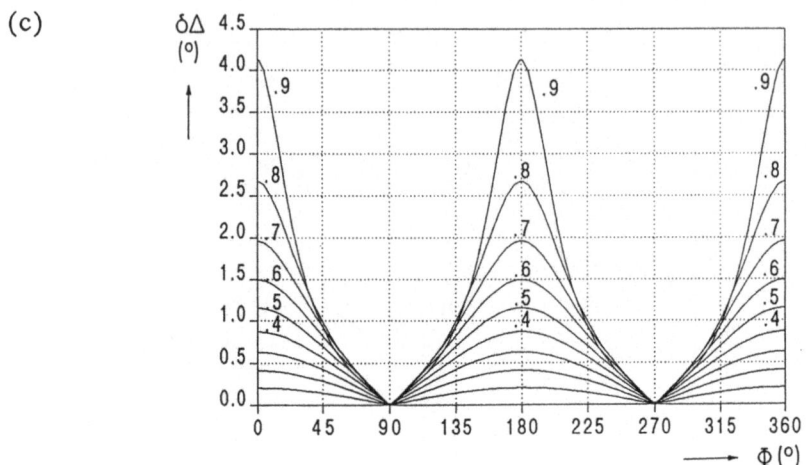

Fig. 33 (c): Uncertainty $\delta\Delta$ as a function of Φ for a phase error $\Delta\Phi = 1°$ and for any polarization of the incident beam; see previous page for a detailed caption.

(a)

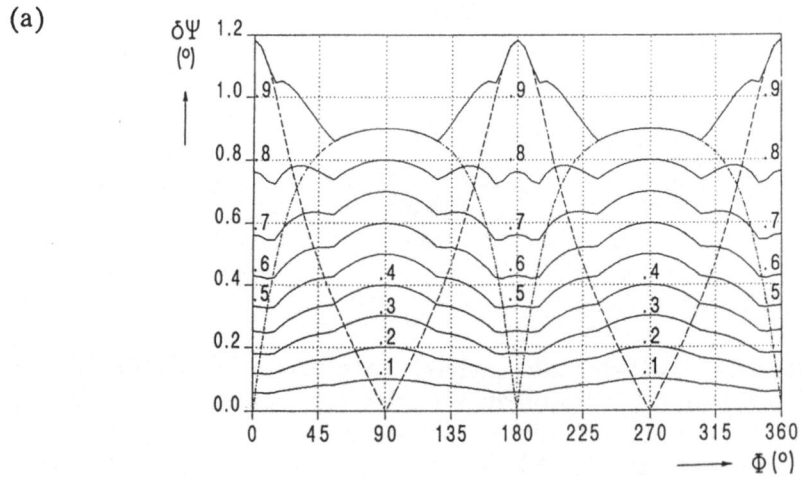

Fig. 34 (a): Uncertainty $\delta\Psi$ as a function of Φ for an amplitude error $\Delta e/e = 0.01$ <u>and</u> a phase error $\Delta\Phi = 1°$ for linearly polarized incident light with $P = 45°$, or for circularly polarized incident light; see next page for a detailed caption.

(b)

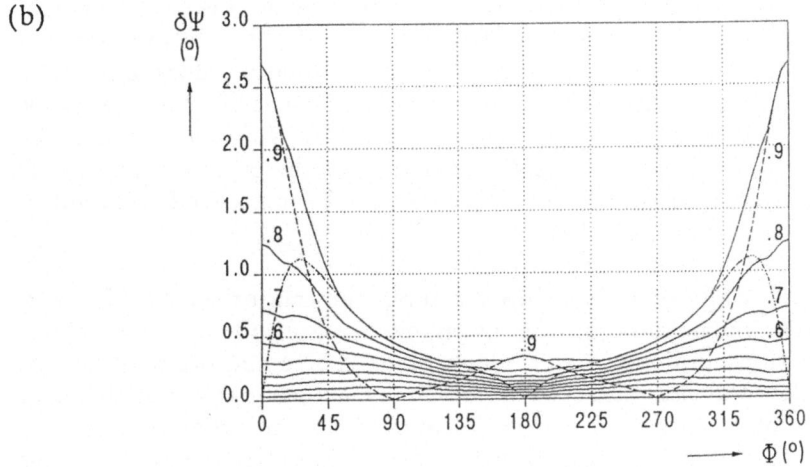

(c)

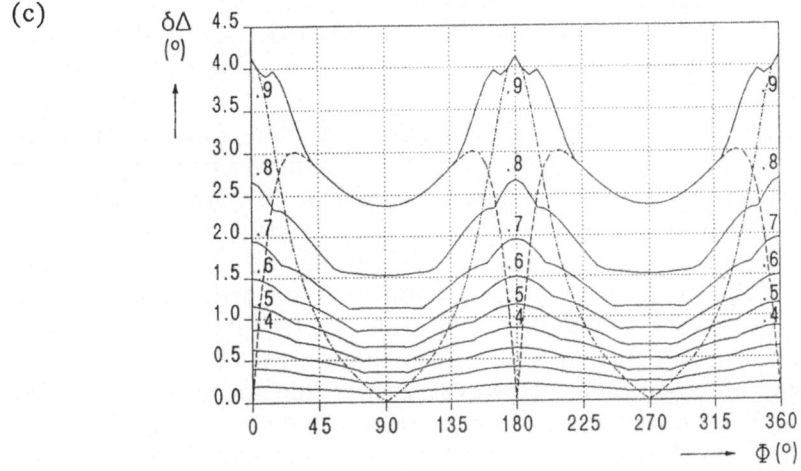

Fig. 34: Rotating analyzer ellipsometry: Uncertainties $\delta\Psi$ (a), (b) and $\delta\Delta$ (c) of the ellipsometric angles for an amplitude error $\Delta e/e$ = 0.01 <u>and</u> a phase error $\Delta\Phi = 1°$ as a function of the phase angle Φ for linearly polarized incident light with P = 45° (a) and P = 15° (b), or for circularly polarized incident light (a); Parameter: normalized AC amplitude e. The graphs for $\delta\Delta$ (c) apply to any polarization of the incident beam. (The normalized amplitude e was varied in steps of 0.1 between 0.1 and 0.9. –––– amplitude error only, –·–·– phase error only. See previous page for Fig. 34 (a).)

In general, the errors increase overproportionately with e. The nar-
rower the polarization ellipse of the reflected beam, the more critical
is the measurement. Due to the degeneration of the relations linking
e and Φ with Ψ and Δ at e = 1, an error estimation does not make
sense altogether for a linearly polarized reflected beam. The errors $\delta\Psi$
and $\delta\Delta$ as functions of Φ are out of phase by 90°, and so are the
errors caused by either Δe or $\Delta\Phi$. This fact might be used to improve
the precision of measurements where one of the measured parameters
is more reliable than the other.

For most rotating analyzer ellipsometers, the major error effect is
probably the uncertainty of the normalized amplitude e: The above
error estimations indicate that a phase error $\Delta\Phi$ in the order of or less
than 10^{-3} degrees can easily be obtained by a reasonable operation of
the instrument, resulting in errors $\delta\Psi$ and $\delta\Delta$ in the order of less than
10^{-3} and $5 \cdot 10^{-3}$ degrees, respectively, if e is kept less than 0.9. The
amplitude error Δe, in contrast, is at least equal to the quantization
error of the analog–to–digital conversion, e.g., $2.5 \cdot 10^{-3}$ for a 12 bit
conversion. Under the same condition – e less than 0.9 –, the uncer-
tainty of Ψ is in the order of $40 \cdot 10^{-3}$ degrees, and the error of Δ, in
the order of $100 \cdot 10^{-3}$ degrees. It should be noted, though, that these
errors are defined as <u>total</u> uncertainties; in the nomenclature of
chapter 3.1, they would be errors of $\pm 20 \cdot 10^{-3}$ and $\pm 50 \cdot 10^{-3}$ degrees for
Ψ and Δ, respectively. The accuracy, at least of the measured Ψ values, can be improved by choosing a polarizer azimuth P which lies in
the vicinity of the expected Ψ angles. Upgrading the 12–bit conversion
which was taken as a base for the above error estimations to the
technically reasonably achievable maximum of 16 bits (resulting in a
resolution of 1/65,536 for e) will reduce these errors by slightly more
than an order of magnitude. However, errors due to noise or encoder
orbiting make themselves felt much more strongly if the intrinsic
accuracy of the system is improved; therefore, the actual reduction of
the uncertainties of Ψ and Δ will probably be less pronounced.

3.2.2.5 Design Rules for Rotating Analyzer Ellipsometers

The results of the above simulations permit the following conclusions:

(1) The intrinsic accuracy of the Fourier system of a rotating analyz-
    er ellipsometer is primarily determined by the resolution of the
    A/D converter used; the DC intensity I_0 and the normalized ampli-
    tude e of the AC component exhibit errors in the order of magni-
    tude of the obtainable resolution of the intensity signal. It is,
    therefore, of utmost importance to keep the amplitude of the light

intensity signal which is submitted to the A/D conversion as close as possible to the maximum which can be processed without an overflow of the A/D converter. The accuracy of the measured phase angle Φ is for all practical cases better than the angle interval between two adjacent sampling points; its error is roughly proportional to the quantization error caused by the finite resolution of the A/D converter.

(2) The number of sampling points per signal period has a smaller influence on the overall accuracy: a greater number of sampling points only helps to reduce errors caused by genuine and quantization noise.

(3) Problems induced by orbiting of the angular encoder for the analyzer position can be overcome very effectively if an even number of signal periods are submitted to the Fourier analysis. For the usual mechanical precision of angular encoders, processing of only one signal period (or of an odd number of signal periods) may result in a significant deterioration of the instrument's accuracy, whereas no noticeable increase over the intrinsic error level ensues from an integration over two periods (or, generally, over an even number of periods). An analysis of the measured intensity signal for the second harmonic may be helpful in identifying potential problems due to encoder orbiting.

(4) Integration over several signal periods is also advisable if problems with periodic noise are suspected. With regard to the suppression of encoder orbiting effects, an even number of periods should be chosen if possible.

(5) The precision of a measurement can be significantly improved by choosing the parameters of the incident light beam (either its polarization azimuth, or, if possible, its ellipticity) such that the reflected beam is as circularly polarized as possible. Although the accuracy of the measured phase angle Φ of the intensity signal declines if the polarization ellipse of the reflected beam approaches a circle, the impact of measurement errors on the ellipsometric angles decreases.

3.3 Effects of the Sample Structure

Even if performed with ultimate accuracy on perfectly aligned (or perfectly numerically calibrated) instruments, ellipsometric measurements may suffer from a third source of error, namely, from an insufficient knowledge of the structure and the parameters of the sample. This effect is felt particularly if the results of an ellipsometric measurement, e.g., the real and imaginary parts of the refractive index of a homogeneous substrate, have to be used as constants in the evaluation of measurements done on a film deposited on the same or an equivalent sample; errors may add up in this case.

3.3.1 Substrate Refractive Index

The influence of a possible uncertainty of ellipsometrically determined substrate parameters was investigated with a similar simulation technique as in the preceding chapters: The "measured" data of an uncoated silicon substrate were calculated and submitted to an error analysis similar to the one presented in chapter 3.1, using the same confidence intervals ($\pm 0.2°$ for Ψ, and $\pm 0.4°$ for Δ). Hence, the substrate refractive index was found to lie between 3.815 and 3.885 with a probability of 95%, and its imaginary part, between -0.008 and -0.032. (The correct values would have been 3.85 and -0.02, respectively; compare Table 1.) The above substrate parameters were, in turn, used for the evaluation of simulated "measured" data generated with the ideal parameters according to Table 1. The extreme film thickness errors x (compare eq. (1) on page 27) and refractive index values obtained from this analysis have been plotted in Fig. 35. The "measured" data were solved for thickness <u>and</u> refractive index in the central part of the ellipsometric order, and for thickness only next to the margins, similar to the approach in chapters 3.1 and 3.2.1.1. The central part of the ellipsometric order turned out to be most sensitive to changes of the real part of the substrate index, while its margins were stronger affected by the imaginary part of the substrate index. In general, the errors match roughly the confidence limits for statistical errors which are indicated as broken lines in Fig. 35.

(a)

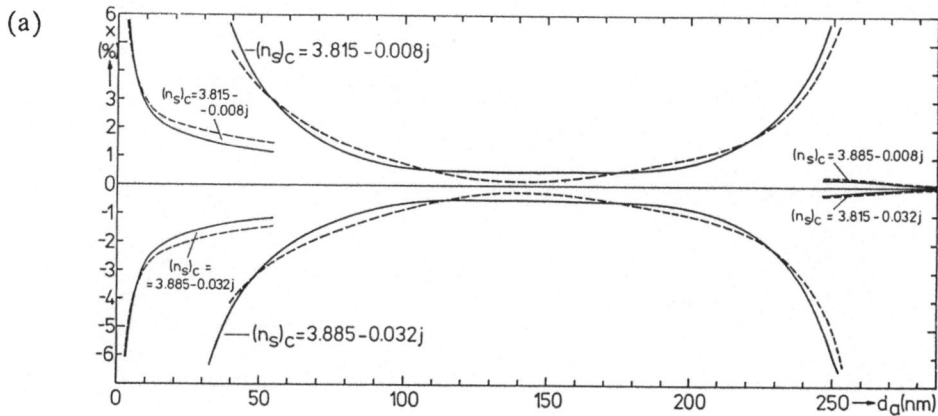

(b)

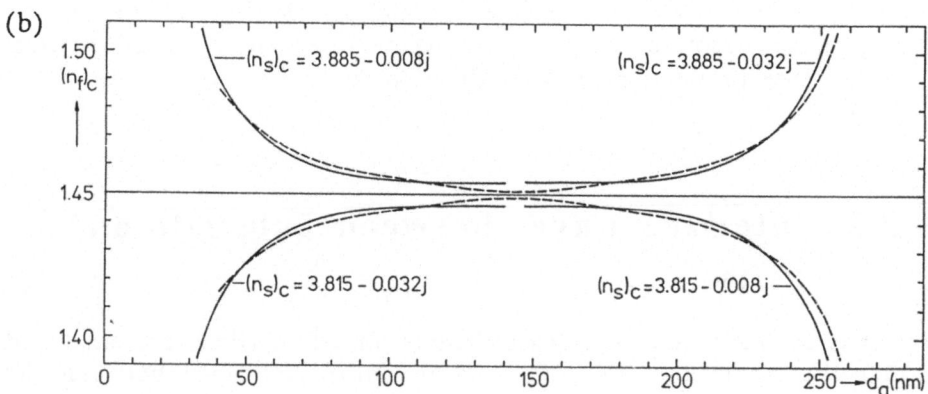

Fig. 35: Substrate data: Influence of the substrate refractive index $(n_s)_c$ used for the calculation of film thicknesses and refractive indices on the relative error x of the computed film thickness (a), on the calculated film refractive index $(n_f)_c$ (b), and on the absolute error of the thicknesses of ultra-thin films (c). (Broken lines: 95% confidence limits for uncertainties $\delta\Psi = 0.2°$ and $\delta\Delta = 0.4°$. See next page for Fig. 35 (c).)

(c)

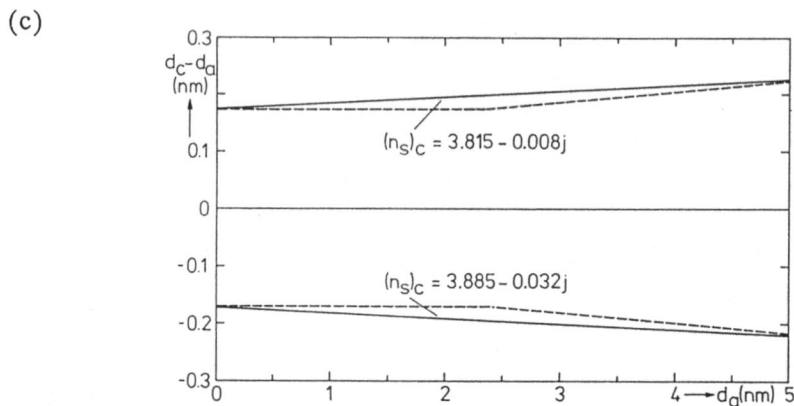

Fig. 35 (c): Substrate data: Influence of the substrate refractive index $(n_s)_c$ used for the calculation of film thicknesses on the absolute error of the thicknesses of ultra–thin films. (See previous page for a detailed caption.)

3.3.2   Interface Layer Between Substrate and Film

In general, ellipsometric measurements on thin films are analyzed under the assumption that there is an abrupt interface between the substrate and the film. However, films need not necessarily start abruptly, particularly if they are grown by some kind of conversion process of the substrate, e.g., by anodic or thermal oxidation, although they may have an abrupt interface even in this case, depending on the preparation technique used [53]. Evidence for transition zones has been reported repeatedly [50],[68]; they may be caused by physical effects like surface roughness [69], or by changes of the chemical composition of the films. Although the thickness of a transition zone is in the order of lattice constants only, it has to be expected to affect measurements at least of ultra–thin films. An obvious cause for an interface layer between the substrate and films deposited by an external process like sputtering or CVD is the native oxide film which grows on many substrates immediately upon contact with air, and which may not have been removed entirely prior to the film deposition.

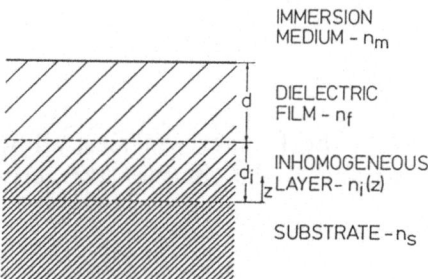

Fig. 36: Model for the analysis of an inhomogeneous interface layer
between the substrate and a dielectric film.

The influence of interface layers was investigated with the model
outlined in Fig. 36: A silicon substrate is assumed to be covered with
an inhomogeneous film whose composition varies between pure silicon
(next to the substrate) and pure silicon dioxide (next to the film).
The interface film is buried under a homogeneous ideal SiO_2 or Si_3N_4
film of arbitrary thickness. Due to the inhomogeneity of the interface
layer, its refractive index n_i will vary between the refractive indices
of pure silicon and SiO_2.

This variation as a function of the distance z from the surface of the
undisturbed substrate was modelled according to two different approaches: The simpler model for inhomogeneous films as provided by
the original McCrackin program [7] assumes a linear variation of the
real part of the refractive index, and a linear variation of the ratio of
the real and imaginary parts of the film index. The second, slightly
more complicated model is based upon a linear variation of the film
composition as a function of the position within the film; it was
implemented in the measurement evaluation program mentioned in
chapter 3.1. In the subsequent function graphs (Figs. 37 – 41), the
results of the first model will be indicated by "a", and those of the
second model, by "b".

There are a number of more or less plausible models for the dependence of the refractive index of a mixed substance on its composition,
based upon the refractive indices of its constituents [9]. Theories
cited in the literature range from a linear interpolation between the
refractive indices of the two components [12],[70] to a linear interpolation of the logarithms of the refractive indices [71]. The only
approach which is founded theoretically (by the law of Clausius and
Mosotti [65],[66]) is based on a linear interpolation, depending on the
volume ratio of the two components, of their "specific refractions"
S(n), where n is the refractive index of the particular component:

$$S(n) = \frac{n^2 - 1}{n^2 + 2} \qquad\qquad (40)$$

This theory was used in the further analyses; allowance was made for complex values of $S(n)$ due to the complex refractive indices involved.

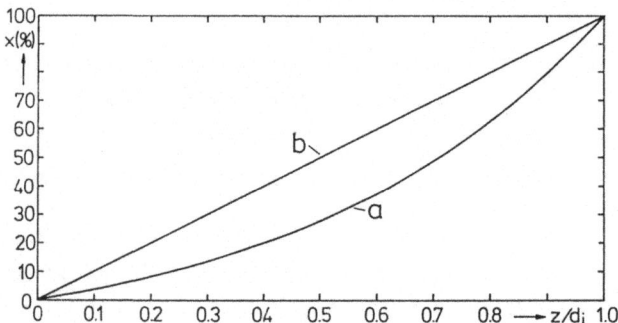

Fig. 37: Variation of the SiO_2 volume content x in an inhomogeneous interface layer between a silicon substrate and an arbitrary film for the two models "a" and "b".

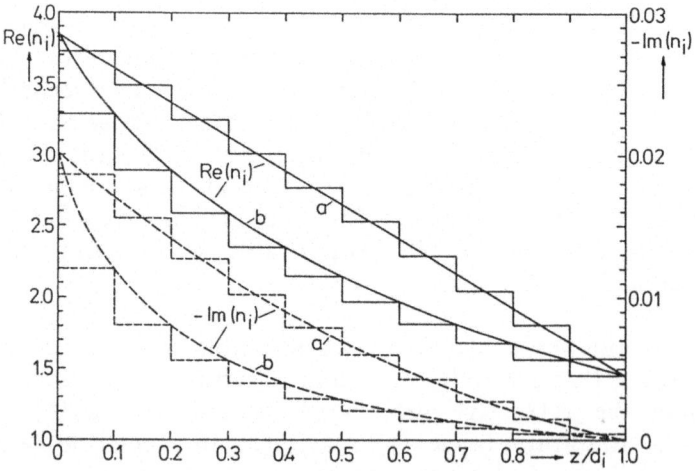

Fig. 38: Variations of the real and imaginary parts of the refractive index of an inhomogeneous film over its thickness d_i according to the models "a" and "b".

"Measured" data were simulated with the two models defined above, for various thicknesses d_i of the inhomogeneous layer. The dependences of the SiO_2 content x in the inhomogeneous film, and of the resulting refractive index n_i, on the distance z from the substrate surface are outlined in Figs. 37 and 38, respectively. For the generation of Ψ–Δ data, the inhomogeneous film was "broken down" into ten sub–layers as indicated by the stepped lines in Fig. 38. From these simulated "measured" data, the thickness and, if feasible, the refractive index of the silicon dioxide or silicon nitride film on top of the inhomogeneous zone were calculated with the standard algorithms for a single homogeneous film.

For the analysis of "measured" data for ultra–thin films (Fig. 39), the correct refractive indices of the silicon substrate and of a homogeneous SiO_2 film were used. The graphs represent the difference between the calculated film thickness d_c (measured between the surfaces of the homogeneous substrate and of the thin film) and the actual distance z of the surface of the SiO_2 film from the substrate surface. The broken parts of the curves correspond to film surface locations which are still within the inhomogeneous layer; they could be interpreted as describing inhomogeneous films which did not yet reach their final thicknesses and/or compositions.

(a)

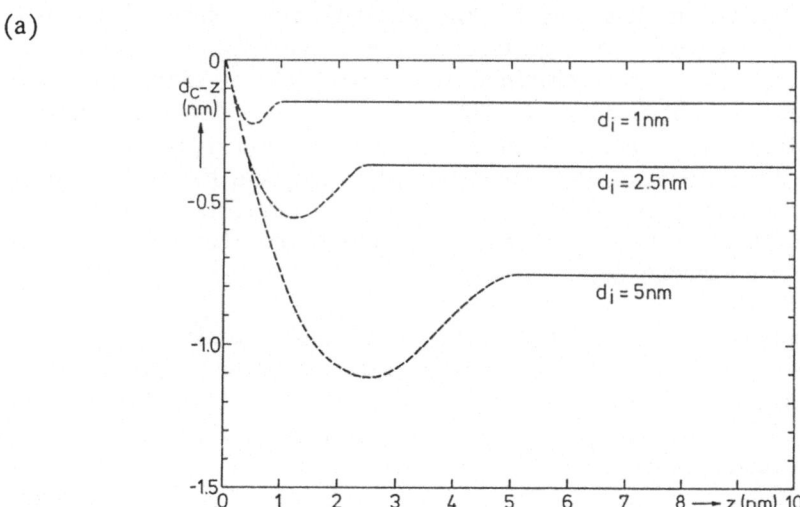

Fig. 39 (a): Inhomogeneous interface layer: Difference between the calculated film thickness d_c and the distance from the substrate for model "a". (See next page for a detailed caption.)

(b)

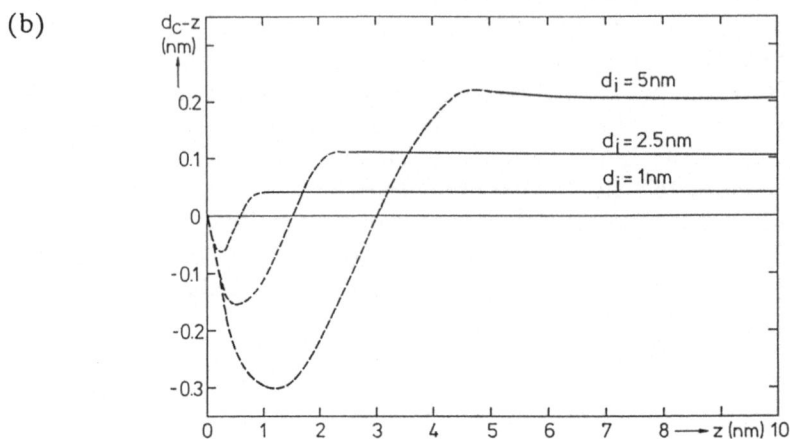

Fig. 39: Inhomogeneous interface layer: Difference between the
          calculated film thickness d_c and the actual distance z be-
          tween the surfaces of the substrate and of the deposited
          ultra–thin SiO_2 film for various thicknesses d_i of the in-
          homogeneous interface layer for the two models "a" (a) and
          "b" (b). (See previous page for Fig. 39 (a).)

Although the two models used for the simulation of measured Ψ and Δ
values do not differ very significantly, there are pronounced differences between the results obtained for them.   These differences may
be attributed to the different subdivision schemes for the inhomogeneous films only to a certain degree; the maximum difference which
could thus be accounted for would be half a stepwidth, i.e., 5% of d_i.
Therefore, not only the thickness but also the microstructure of an –
albeit ultra–thin – inhomogeneous film seems to significantly influence
its effect on measured data.

Inhomogeneous films – in the form of a native oxide layer – may
already have been present when the refractive index of the substrate
was determined ellipsometrically.   Hence, the above evaluation was
repeated with substrate data computed from a "measurement" of a
substrate already covered with an inhomogeneous film. The analysis of
ultra–thin films subsequently "deposited" on top of the inhomogeneous
layer always resulted in the correct thicknesses of the dielectric films.
This result may appear trivial but it is not: For the correct analysis
of a multi–layer structure, the reflection coefficients for light polarized in parallel and perpendicular to the plane of incidence have to be
calculated for each interface, starting with the surface of the sub-

strate and introducing the overlying layers one by one, combining them with the underlying layers to a fictive substrate (compare chapter 1.4). Finally, the complex reflection coefficient ρ (or, equivalently, Ψ and Δ) is calculated from the quotient of the two reflection coefficients at the top surface for the two orthogonal states of polarization. Since either of these reflection coefficients is a complex quantity, the behavior of the sample is described by four parameters up to that point (namely, by the real and imaginary parts of each of the two reflection coefficients). Calculating the refractive index of a fictive substrate entails, however, the loss of two of these four parameters; films on top of such a fictive substrate may or may not be evaluated correctly therefore.

These considerations could be proved in the next simulation step where a thicker SiO_2 or Si_3N_4 film was considered deposited on top of a silicon substrate covered by a 2.5 nm thick inhomogeneous layer. The refractive index of this fictive substrate was determined and used for the analysis of the film data; these fictive substrate indices depended slightly on the model used ($3.840-j\cdot0.17$ for model "a", and $3.837-j\cdot0.20$ for model "b"). The simulated ellipsometric angles were solved for film thickness d_c and refractive index $(n_f)_c$ for an entire ellipsometric order (0 – 286.5 nm for SiO_2, and 0 – 179.2 nm for Si_3N_4); the results are shown in Fig. 40.

The above analysis was, finally, repeated with the parameters of an ideal uncoated silicon substrate. The results in Fig. 41 indicate a shift of the calculated thickness which is about equal to the thickness d_i of the inhomogeneous film. This is plausible since the thickness of the entire film is now measured from the surface of the substrate, including the inhomogeneous layer. In contrast to the previous analysis, there is only a small but virtually constant difference in the thickness readings for both models which matches the difference seen for ultra–thin films (compare Fig. 39).

The results of the above simulations indicate therefore that there is a not entirely negligible influence of inhomogeneous interface layers on the thickness and, particularly, on the refractive index determined for films of arbitrary thickness. The deviations of the measured refractive index from its true value, especially next to the boundaries of a thickness multiple, are liable to deteriorate the accuracy of thickness measurements of films several ellipsometric orders thick (compare chapter 3.1). For precision measurements, suitable compensation for the influence of interface layers should therefore be attempted if their existence is suspected.

(a)
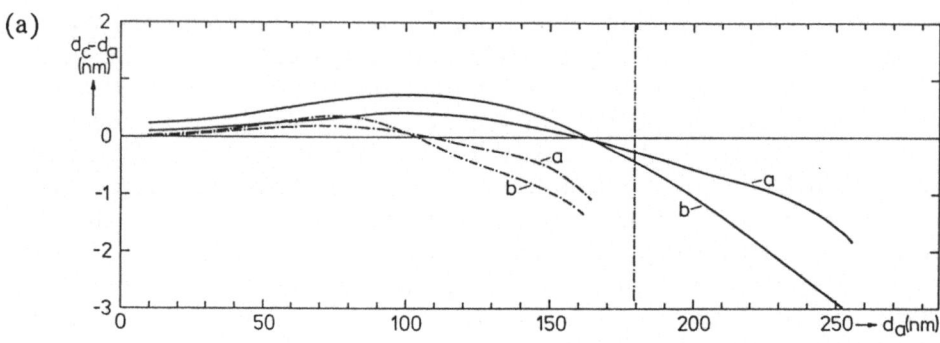

(b)
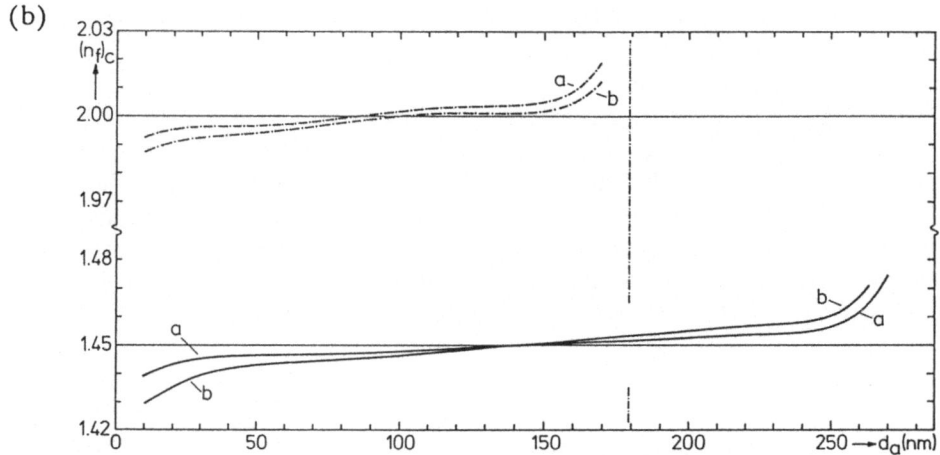

Fig. 40: Inhomogeneous interface layer: Absolute errors of the calcu‐
lated thickness d_c (a) and refractive index $(n_f)_c$ (b) over the
first ellipsometric order for the two models "a" and "b".
Calculation based on fictive substrate refractive indices.
(Full lines: SiO_2 films, broken lines: Si_3N_4 films.)

(a)

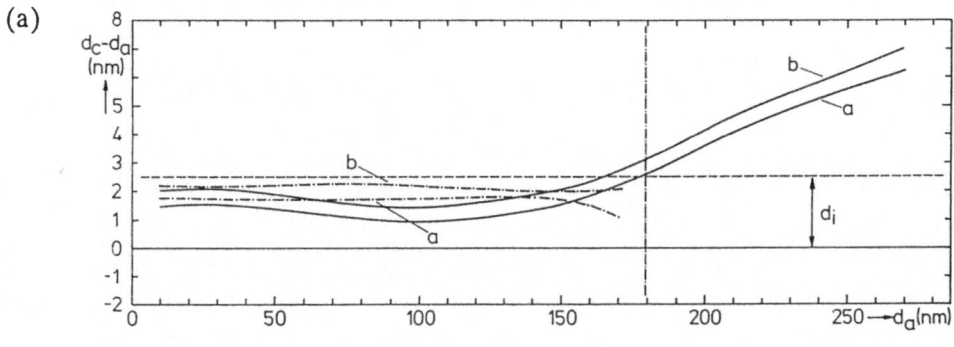

(b)

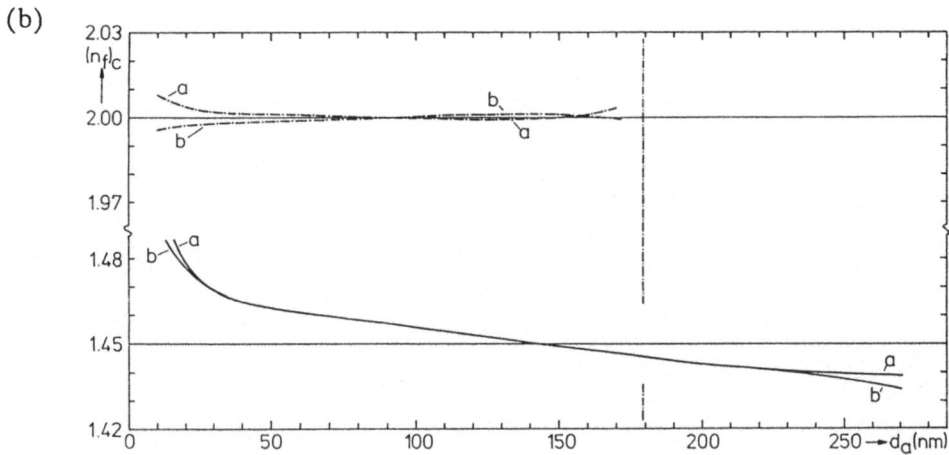

Fig. 41: Inhomogeneous interface layer: Absolute errors of the calculated thickness d_c (a) and refractive index $(n_f)_c$ (b) over the
first ellipsometric order for the two models "a" and "b".
Calculation based on the actual substrate refractive index.
(Full lines: SiO_2 films, broken lines: Si_3N_4 films.)

Appendix A: Design Considerations for a High-Speed Rotating Analyzer Ellipsometer

Ellipsometry as a non–invasive and non–destructive optical technique is particularly suited for the *in situ* observation of processes, e.g., of the thin film deposition and etching processes involved in the production of microelectronic components. A venture was therefore launched to implement ellipsometry as a monitoring technique for various plasma-based film deposition and etching processes which are being developed at the Technical University of Vienna. Since no suitable instrument was commercially available when the project was planned (and even the *in situ* ellipsometer which is on the market now is not fast enough for the monitoring of critical process stages; compare chapter 1.4), an instrument had to be specially constructed. With regard to the superior data acquisition speed of the rotating analyzer approach, the instrument was based on this principle. For reasons of simplicity, and in order to avoid errors introduced by a retarder, the instrument was designed for operation with a linearly polarized incident probe beam only, i.e., without a quarter–wave plate after the polarizer prism. The instrument is currently still under construction; no actual performance data or results of measurements done with it can therefore be given yet.

The construction of a high–speed ellipsometer entails the development of three major sub–systems which will be discussed below in more detail:

(1) An optical assembly, consisting of the light source, the polarizer and analyzer prisms in suitable rotating stages, the photodetector, and various auxiliary components like diaphragms and filters.

(2) Electronic circuitry which controls the various adjustable components, and conditions the measured signals for further processing.

(3) A sufficiently efficient microcomputer which has to process the measured data and to provide control output signals for the interface circuitry.

A1 The Optical Assembly

The optical assembly of the *in situ* ellipsometer is outlined schematically in Fig. A1.

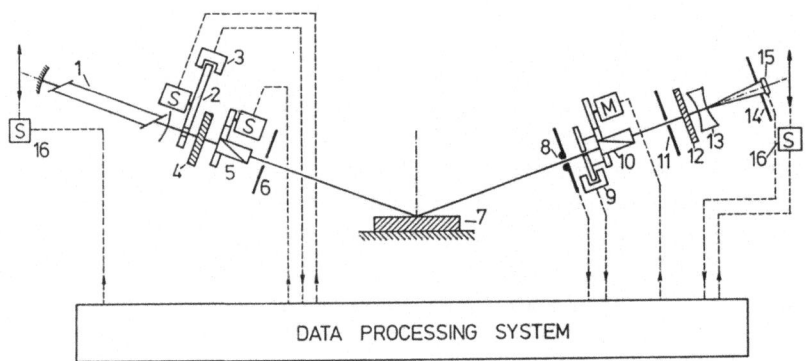

Fig. A1: Optical assembly of a high–speed rotating analyzer ellipsometer:

1	He–Ne laser.	9	Analyzer encoder.
2	Shutter.	10	Analyzer prism.
3	Shutter encoder.	11	Diaphragm.
4	λ/4 plate.	12	Bandpass filter.
5	Polarizer prism.	13	Dispersing lens.
6	Diaphragm.	14	Diaphragm.
7	Sample.	15	Photodetector.
8	Beam sensors.	16	Alignment system.

Because of the high data acquisition speed planned, spectroscopic measurements did not appear practical. The ellipsometer was therefore designed for operation at only one wavelength; with regard to the relatively high intensity obtainable, a low–noise helium–neon laser (1) with a wavelength of 632.8 nanometers was chosen as a light source. Light emitted by the laser passes through a chopper/shutter system (2) which permits to shut off the laser beam under control of the data processing system. (This is necessary for a compensation of the influence of stray light, of background light emitted by a plasma process monitored, and of the dark current of the photodetector. The intensity measured with the shutter closed can simply be subtracted from the intensity obtained with the shutter open, provided that the background signal is constant and not too noisy.) The position of this shutter – essentially a disk with cutouts – is monitored by means of an optical encoder (3). The light emitted by the laser is linearly polarized; the optical noise level of an unpolarized laser would have been too high. In order to permit an approximately constant intensity of the probe beam independent of the orientation of the polarizer, the linearly polarized laser beam is converted to circular polarization by means of a quarter-wave plate (4). Since this function was considered necessary but not critical, a simple foil-based device was chosen for this purpose. The polarizer (5) which follows immediately after the de-polarization wave plate is a Glan-Thompson prism with broad-band antireflection coatings; it is mounted in a precision rotary stage moved by a stepping motor. The polarizer azimuth need not be adjusted frequently and quickly; a commercial rotary stage could therefore be used. A diaphragm (6), finally, limits the beam diameter and cuts off possible parasitic beams.

After its reflection at the sample (7), the beam passes through a second diaphragm (8) which is, in addition, equipped with photodetectors which are used to sense the beam location during the initial alignment of the instrument. A special rotary stage holds the analyzer Glan-Thompson prism (10) whose design is identical to the one used for the polarizer. The azimuth of the rotating analyzer is measured with an incremental encoder with 1024 divisions (9).

The construction of the analyzer rotary stage caused major problems: The duration of a measurement was intended to be less than 10 milliseconds in order to permit a good resolution of fast processes. Since a measurement can be done in one signal period, i.e., during half a rotation of the analyzer (although two signal periods would be preferable with regard to accuracy and noise suppression; compare chapter 3.2.2), an analyzer rotation speed in excess of 3000 rpm is required. Speeds up to 6000 rpm were taken as a base for the design of the electronic circuitry, but the mechanical limitations of the DC motor used for rotating the analyzer stage restricted its speed to 5000 rpm. The first analyzer assembly consisted of a prism holder coaxially mounted on the motor shaft and running on ball bearings. Although

the prism holder was designed with a relatively large mass to guarantee its smooth motion due to its mechanical inertia, the analyzer
assembly proved to be unusable due to the vibrations and torque
fluctuations caused by the ball bearings. It had, accordingly, to be
re-designed with an air bearing for the critical prism holder.

Having passed through a third diaphragm (11), the beam is directed
through an optical bandpass filter (12) which reduces the influence of
broad-band stray light. A dispersing lens (13) widens the beam in
order to avoid disturbances caused by a possible orbiting of the beam
on the photodetector (15). ("Blind spots" on the detector which can
very easily originate from dust particles have a stronger impact on the
measurement if their size is in the order of magnitude of the beam
diameter, and if the beam orbits over them.)

A simple silicon photodiode with an active area of about 10 mm^2 is
used as a photodetector. Estimates of the light intensity arriving at
the detector showed that the sensitivity of such a photodiode is by far
sufficient. Its photocurrent depends only on the total light intensity
and on the quantum efficiency but not on the area of the diode.
Since the quantum efficiencies of a variety of diodes evaluated are
comparable, the choice was, finally, based on the specified dark cur-
rents. The speed of the detector is rather irrelevant since the frequency of the AC component of the signal is less than 200 Hz even
for the maximum possible rotation speeds.

An alignment system (16) was planned for the improvement of the
accuracy of the ellipsometer under genuine *in situ* measurement conditions. The error analyses of chapter 3.2.1.1 indicate that even a minor
error of the angle of incidence of the probe beam can very easily
result in an insufficient accuracy of the measurements although the
beam is still able to pass through the diaphragms of the analyzer arm.
For reasons of simplicity, the instrument is designed for a fixed angle
of incidence of 70°; the polarizer and analyzer arms are mounted
under a fixed angle of 140° accordingly. In advanced conventional
systems, the sample holder can be adjusted for maximum intensity at
the detector in order to eliminate the influence of a possible wedge
shape of the sample. This is obviously not possible if the sample is
kept within some kind of film deposition or etching equipment; the
compensation of sample thickness fluctuations and wedge angles has
therefore to be done by moving the entire ellipsometer. The motion
of the two spindles which permit to raise or lower either side of the
instrument can be controlled by the data processing system according
to the intensity information obtained from the main photodetector and
the auxiliary photodiodes on diaphragm (8).

A2 Electronic Interface Circuitry

The circuitry which directly controls the operation of the instrument and which pre-processes the measured signals for further evaluation is outlined in Fig. A2 together with the block diagram of the data acquisition computer which will be discussed in the next chapter. The interface comprises controllers for the DC motor of the analyzer assembly and for the stepping motors for the shutter, the polarizer rotary stage, and the alignment system. In addition, it contains the pre-amplifiers for the main and the auxiliary photodetectors, and pulse-shaping circuitry for the signals generated by the incremental encoder on the analyzer shaft. All these circuits can be controlled and programmed by the data acquisition microcomputer.

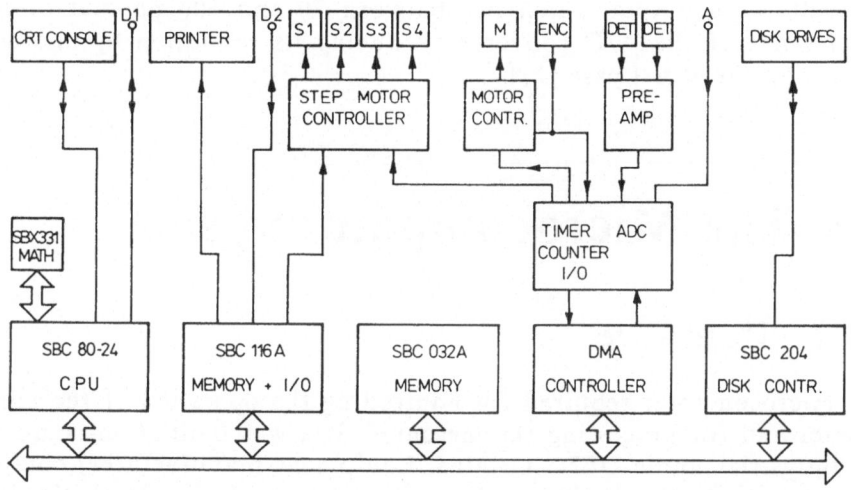

Fig. A2: Data processing system of the high-speed rotating analyzer ellipsometer. (D1, D2 – digital I/O; S1 ... S4 – stepping motors; M – analyzer DC motor; ENC – analyzer encoder; DET – photodetectors; A – auxiliary analog input.)

While the design of the stepping motor drivers was relatively straight-forward, a major effort was needed to guarantee an accurate operation of the analyzer DC motor. This is particularly important since the rotation speed directly affects the compensation of delays and phase shifts caused by the analog photodetector circuitry [16],[17]. A phase-locked loop control approach was therefore chosen where a signal derived from the incremental encoder is compared to a clock signal either supplied by a timer within the microcomputer system, or by a dedicated crystal oscillator in the motor controller [72].

The design of the pre-amplifier circuitry for the photodetector signal is not straightforward either: The error estimation of chapter 3.2.2 demands that the intensity of the signal supplied to the analog-to-digital (A/D) converter should lie as close as possible to the maximum permitted by the converter, in order to guarantee a sufficient resolution of the conversion.   Since the average light intensity on the detector may vary over a wide range, depending on the sample and operation parameters, this implies a variable gain of the pre-amplifier. However, the total phase shift of the analog circuitry must remain constant independent of the gain in order to permit its compensation. A phase compensated amplifier was therefore designed which permits to select its gain in 18 steps on a square-root of two scale.   Hence, the gain of the preamplifier can be set for any arbitrary signal level to result in an output which lies between 70 and 100 percent of the signal range of the A/D converter, reducing its resolution by one half bit only in the worst case [73].

A3 The Microcomputer System

A3.1 Hardware

The microcomputer required for controlling the operation of the ellipsometer and for processing the measured data was built from commercial industrial-grade OEM (Original Equipment Manufacturer) components based on the Intel Multibus system.   Its design is outlined in Fig. A2:   A main processor board (Intel iSBC 80-24) holds the 8-bit 8085-AH2 CPU (Central Processing Unit) operated at 5 MHz, two banks of 8 KBytes ROM (Read Only Memory) each, 8 KBytes of static RAM (Random Access Memory), one serial port (which is connected to the console CRT terminal) and 24 digital input/output lines.   An iSBX 331 board rides piggy-back on the main processor board; it contains an Intel 8231A APU (Arithmetic Processing Unit) which provides the high-speed integer and floating-point arithmetic which is required for the Fourier transformation of the measured data.   An Intel iSBC 116A and an Intel iSBC 032A board complete the addressable memory space of 64 KBytes with 16 and 32 KBytes of RAM, respectively; in addition, they provide one serial interface (to which a matrix printer is connected) and 24 parallel I/O lines. An Intel iSBC 204 flexible disk controller board, finally, supervises two industrial standard 8" flexible disk drives.

A DMA (Direct Memory Access) controller board (actually, a set of two boards) was specially designed as an interface to the ellipsometer circuitry. The ellipsometer was planned for taking up to 1024 data points per half–rotation of the analyzer, submitting them to A/D conversion, and applying Fourier algorithms for obtaining the amplitude and phase of the intensity fluctuations at the photodetector. (A fraction of this number of sampling points per period can be chosen under software control if the data evaluation time is critical. Up to 64 sampling points per period can be processed by the Fourier analysis routines in a real–time mode, i.e., while the next set of sampling points is being input.) If the analyzer is rotating at its full design speed of 6000 rpm, the interval between two adjacent data sampling points will therefore be in the order of 5 μs; data arriving at this rate cannot be handled by the CPU any more and have to be deposited in the microcomputer's memory by direct memory access without intervention of the CPU.

Two Multibus–compatible boards were designed to combine the functions of the DMA controller (which is a standard VLSI circuit), the necessary bus arbitration logic, the A/D converter with its supporting circuitry, and the multiplexers and the sample/hold amplifier for the analog input. With regard to the high speed requirements imposed on data acquisition, a 12 bit A/D converter with a conversion time of 3 μs had to be chosen. In addition, the DMA boards were designed to hold two timer chips with three timers each, and one parallel interface circuit with 24 digital I/O lines.

A3.2 Software

Controlling an instrument like an ellipsometer, and processing the measured data on–line, with the target of controlling a technological process, is, obviously a real–time computer application which requires special efforts for the synchronization of the microcomputer's operations to external events. For relatively complex software like the one conceived for the ellipsometer, the use of a dedicated real–time operating system offers significant advantages over the standard (however, more efficient) approach of entirely custom–made interrupt driven software. Hence, Intel's Real Time Executive for 8080/85 microprocessors, iRMX–80 [74], was chosen as an operating system for the data acquisition microcomputer. Several functions of iRMX–80 were improved or extended by custom–written system routines, e.g., an enhanced terminal and printer interface, and a generic loader routine for program code. The real–time environment provided by iRMX–80 was further augmented by specially designed emulation software for Intel's System Implementation Supervisor operating system ISIS–II [75], which permits to run auxiliary programs (e.g., for disk or file maintenance) on the ellipsometer's computer. The operating system and all application software is loaded from disk; only a small nucleus of the operating system is permanently kept in ROM. This design did not only facilitate the development and implementation of measurement evaluation software, it also greatly enhanced the flexibility of the controller computer which can be regarded, indeed, as a general purpose process controller.

The software used for the evaluation of the measured intensity data of the ellipsometer has to meet two – partly opposing – requirements, namely, high accuracy and high execution speed. Particularly the Fourier transformation of the intensity samples affects critically the performance of the ellipsometer because it has to be executed "on-line" during a measurement in any case in order to concentrate the measured data into a manageable amount, even if it is possible to evaluate these data "off–line" after an experiment. Therefore, the following approach was chosen:

(1) The sine and cosine coefficients required for the Fourier transformation are kept in a table rather than being calculated again and again. Indeed, it is sufficient to provide only one quadrant of the sine function

$$\sin (2\pi\nu/n) \text{ for } 0 \leq \nu \leq n/4$$

in the table which can be read up and down as required to give the correct sine or cosine coefficients. With regard to the maximum number of sampling points per period, 1024, the sine table has to hold 257 entries (256 + 1 to cover both quadrant boundaries).

(2) The Fourier coefficients are, in general, calculated with integer rather than floating–point algorithms. The APU used supports 16 and 32 bit signed integer and 32 bit floating–point operations (the latter with 23 bit mantissa accuracy, which is a standard for 8 bit microcomputer systems). The execution time increases significantly in the order given. The most time–efficient approach which preserves the full resolution of the A/D converter output, 12 bits,

was found as follows: The 16 bit sine coefficients are multiplied with the 12-bit intensity values using a 16 bit algorithm which returns the most significant 16 bits of the 32 bit product; the sums are accumulated in 32 bit integer words. Only the final steps, the scaling and the transfer from the APU to the microcomputer proper, are done in floating-point notation. Optimized programming of the chosen hardware resulted in a total processing time of about 140 μs per sampling point, thus permitting the handling of approximately 7000 sampling points per second.

(3) The output data generated by the 12-bit A/D converter which range originally between 0 and 4095 are multiplied by 8 (i.e., left-shifted by three bits) by the A/D converter hardware prior to being deposited in the microcomputer's memory via DMA. The information contained in the least significant bit can thus be retained; otherwise, it would have been lost when the Fourier products are calculated.

References

1. Azzam, R.M.A., Bashara, N.M.: Ellipsometry and polarized light. Amsterdam: North Holland. 1977.
2. Zaininger, K.H., Revesz, A.G.: Ellipsometry – a valuable tool in surface research. RCA Rev., 25, 85 (1964).
3. Aspnes, D.E.: Studies of surface, thin film and interface properties by automatic spectroscopic ellipsometry. J. Vac. Sci. and Technol., 18, 289 (1981).
4. Azzam, R.M.A.: Ellipsometric configurations and techniques. Proc. Soc. Photo–Opt. Instrum. Eng., 276, 180 (1981).
5. Schaefer, R.R.: Infrared ellipsometry – a new technique for characterization of dopant parameters in silicon. J. de Phys., C10, 87 (1983).
6. Braun, P., Störi, H., Söllner, E.: Verfahren zur Messung wetterbedingter Zustandsänderungen an der Oberfläche von Verkehrsflächen und Vorrichtung zur Durchführung des Verfahrens. Application No 84904064.7 at the European Patent Office, 1983.
7. McCrackin, F.L.: A Fortran program for analysis of ellipsometer measurements. Natl. Bur. Stand. (U.S.), Techn. Note 479 (1969).
8. Riedling, K.: Error effects in the ellipsometric investigation of thin films. Thin Solid Films, 75, 355 (1981).
9. Riedling, K.: Technologie legierter Planardioden aus Indiumantimonid. Doctor's Thesis, Vienna: TU Wien. 1979.
10. Belinska, A.A., Kalnynya, R.P., Feltyn, I.A.: Ellipsometric studies of highly absorbing films on the surface of silicon. Opt. and Spectrosc., 46, 175 (1979).
11. Nakata, J., Kajiyama, K.: The anomalous refractive index in the ellipsometric evaluation of an inhomogeneous film. Thin Solid Films, 80, 383 (1981).
12. Applied Materials: Instruction manual for Ellipsometer II. Santa Clara, Calif.: Applied Materials. 1976.
13. Alterovitz, S.A., Bu–Abbud, G.H., Woollam, J.A., Liu, D.C.: An enhanced sensitivity null ellipsometry technique for studying films on substrates: Application to silicon nitride on gallium arsenide. J. Appl. Phys., 54, 1559 (1983).
14. Riedling, K.: Evaluation of adjustment data for simple ellipsometers. Thin Solid Films, 61, 335 (1979).

15. Drevillon, B., Perrin, J., Marbot, R., Violet, A., Dalby, J.L.: Fast polarization modulated ellipsometer using a microprocessor system for digital Fourier analysis. Rev. Sci. Instrum., 53, 969 (1982).

16. Aspnes, D.E.: Fourier transform detection system for rotating-analyzer ellipsometers. Opt. Commun., 8, 222 (1973).

17. Aspnes, D.E.: Optimizing precision of rotating–analyzer ellipsometers. J. Opt. Soc. Amer., 64, 639 (1974).

18. Aspnes, D.E., Studna, A.A.: High precision scanning ellipsometer. Appl. Optics, 14, 220 (1975).

19. Chandler–Horowitz, D.: Ellipsometric accuracy and the principal angle of incidence. Proc. SPIE, 342, 121 (1982).

20. Chandler–Horowitz, D., Candela, G.A.: Principal angle spectroscopic ellipsometry utilizing a rotating analyzer. Appl. Opt., 21, 2972 (1982).

21. Aspnes, D.E.: Effects of component optical activity in data reduction and calibration of rotating–analyzer ellipsometers. J. Opt. Soc. Amer., 64, 812 (1974).

22. Gaertner Scientific Corporation: Auto Gain Ellipsometers. Bulletin EE, Chicago, Illinois: Gaertner Scientific Corp. 1987.

23. Fresnel, A.: Ann. Chim. et Phys., 28, 147 (1825).

24. Rudolph Research: Data sheet on Automatic Ellipsometer AutoEl–II. Techn. Bull. 461A, Fairfield, New Jersey: Rudolph Research. ca. 1980.

25. Kutko, R.J.: Ellipsometry for semiconductor process control. Solid State Technol., 21, 2, 43 (1978).

26. Vina, L., Cardona, M.: Optical constants of pure and heavily doped silicon and germanium: electronic interband transitions. Proc. 16th Intern. Conf. on the Phys. of Semicond., Montpellier. 1982.

27. Aspnes, D.E.: Optical characterization by ellipsometry – a prospective. J. de Phys., C10, 3 (1983).

28. Theeten, J.B., Chang, R.P.H., Aspnes, D.E., Adams, T.E.: *In situ* measurement and analysis of plasma–grown GaAs oxides with spectroscopic ellipsometry. J. Electrochem. Soc., 127, 378 (1980).

29. Aspnes D.E. et al.: Optical properties of GaAs and its electrochemically grown anodic oxide from 1.5 to 6 eV. J. Electrochem. Soc. 128, 590 (1981).

30. Aspnes, D.E., Studna, A.A.: Optical detection and minimization of surface overlayers on semiconductors using spectroscopic ellipsometry. Proc. Soc. Photo–Opt. Instr. Eng., 276, 227 (1981).

31. Theeten J.B. et al.: Nondestructive analysis of $Si_3N_4/SiO_2/Si$ structures using spectroscopic ellipsometry. J. Appl. Phys., 52, 6788 (1981).

32. Bilenko, D.I., Dvorkin, B.A.: Laser ellipsometer for the investigation of transient processes. Instrum. and Exp. Tech., 23, 188 (1980).

33. Greef, R.: Ellipsometry. 11. Bunsen–Kolloquium, Vienna. 1982.
34. Theeten, J.B., Hottier, F., Hallais, J.: On–time determination of the composition of III–V ternary layers during VPE growth. Appl. Phys. Lett., 32, 576 (1978).
35. Theeten, J.B., Hottier, F., Hallais, J.: Ellipsometric assessment of (Ga,Al)As/GaAs epitaxial layers during their growth in an organo-metallic VPE system. J. Cryst. Growth, 46, 245 (1979).
36. Laurence, G., Hottier, F., Hallais, J.: Growth monitoring and characterization of (Al,Ga)As heterostructures by ellipsometry. J. Cryst. Growth, 55, 198 (1981).
37. Hottier, F., Cardoret, R.: *In situ* observation of polysilicon nucle-ation and growth. J. Cryst. Growth, 56, 304 (1982).
38. Busta, H.H., Lajos, R.E.: Ellipsometric end point detection during plasma etching. Proc. 1977 Intern. Electron. Dev. Meeting, Wash-ington, DC. 1977.
39. Theeten, J.B., Chang, R.P.H., Adams, T.E.: *In situ* analysis of plasma–grown oxides using a spectroscopic ellipsometer. J. Vac. Sci. and Technol., 16, 216 (1979).
40. Theeten, J.B.: Real–time and spectroscopic ellipsometry of film growth: Application to multilayer systems in plasma and CVD processing of semiconductors. Surf. Sci., 96, 275 (1980).
41. Bilenko, D.I., Galishnikova, Y.N., Dvorkin, B.A.: Ellipsometric monitoring of dielectric layers in semiconductor structures during their preparation. Opt. and Spectrosc., 45, 58 (1978).
42. Riedling, K.: Dynamische *in situ*–Ellipsometrie für Grundlagenun-tersuchungen und Prozesskontrolle. Fresenius Z. Anal. Chem., 319, 706 (1984).
43. Perrin, J., Drevillon, B.: Growth characterization of Si:H films by multiple angle of incidence spectroscopic ellipsometry. J. de Phys., C10, 247 (1983).
44. Erman, M., Theeten, J.B.: Multilayer analysis of ion implanted GaAs using spectroscopic ellipsometry. Surf. and Interf. Analys., 4, 98 (1982).
45. Ohira, F.: Ellipsometric characterization of Si surface damage induced by sputter etching. J. Electrochem. Soc., 130, 1201 (1983).
46. Ged, P., Debroux, M.H.: Optical evidence of structural modification of implanted silicon during furnace annealing. J. de Phys., C10, 267 (1983).
47. Lohner, T., Mezey, G., Kotai, E., Paszti, F., Manuaba, A., Gyulai, J.: Characterization of ion–implanted silicon by ellipsometry and channeling. Nucl. Instr. and Methods, 209/210, 615 (1983).
48. Delfino, M., Razouk, R.R.: Optical constants of arsenic and boron implants in silicon determined by a four–phase complex refractive index model. J. Electrochem. Soc., 129, 606 (1982).

49. Hottier, F., Hallais, J., Simondet, F.: *In situ* monitoring by ellipsometry of metalorganic epitaxy of GaAlAs–GaAs superlattice. J. Appl. Phys., 51, 1599 (1980).

50. Palik, E.D., Bermudez, V.M.: Ellipsometric investigation of the silicon/anodic oxide interface. J. de Phys., C10, 179 (1983).

51. van de Ven, E.P.G.T.: Plasma deposition of silicon dioxide and silicon nitride films. Solid State Technol., 24, 4, 167 (1981).

52. Hollahan, J.R., Bell, A.T.: Techniques and applications of plasma chemistry. New York: J. Wiley. 1974.

53. Aspnes, D.E., Theeten, J.B.: Spectroscopic analysis of the interface between Si and its thermally grown oxide. J. Electrochem. Soc., 127, 1359 (1980).

54. Buckman, A.B., Chao, S.: Ellipsometric characterization of the glassy layer at metal/semiconductor interfaces. Proc. 4th Intern. Conf. on Ellipsometry, Berkeley, Calif., 1979.

55. Bootsma, G.A.: Characterization of interfaces. Sens. and Act., 1, 289 (1981).

56. Candela G.A. et al.: Measurement of the interlayer between aluminum and silicon dioxide using ellipsometric, capacitance–voltage, and Auger electron spectroscopy techniques. Thin Solid Films, 82, 183 (1981).

57. Donnelly, V.M., Flamm, D.L.: Anisotropic etching in chlorine-containing plasmas. Solid State Technol., 24, 4, 161 (1981).

58. Moran, J.M.: High resolution resist patterning using reactive ion etching techniques. Solid State Technol., 24, 4, 195 (1981).

59. Lam, D.K.: Advances in VLSI plasma etching. Solid State Technol., 25, 4, 215 (1982).

60. Riedling, K.: Accuracy of digital Fourier transformation detection systems for high speed rotating analyzer ellipsometers. Thin Solid Films, 155, 151 (1987).

61. Chandler–Horowitz, D., Candela, G.A.: On the accuracy of ellipsometric thickness determination for very thin films. J. de Phys., C10, 23 (1983).

62. McCrackin, F.L., Passaglia, E., Stromberg, R.R., Steinberg, H.: Measurement of the thickness and refractive index of very thin films and the optical properties of surfaces by ellipsometry. J. Res. Nat. Bur. Stds., 67A, 363 (1963).

63. Przyborski, W., Roed, J., Lippert, J., Sarholt–Kristensen, L.: A refined oxidation–stripping technique of thin n–type Si films. Rad. Eff. 1, 33 (1969).

64. Grovenor, C.R.M.: Atom probe microanalysis. IBM Europe Institute, Oberlech, Austria. 1986.

65. Pohl, R.W.: Optik. Berlin: Springer. 1940.

66. Hund, F.: Theoretische Physik, Vol. 2. Stuttgart: B.G. Teubner. 1957.

67. Knuth, D.E.: The art of computer programming, Vol. 2 (Semi-numerical algorithms). Reading, Mass.: Addison–Wesley. 1969.
68. McFeely, F.R.: Core level spectroscopy. IBM Europe Institute, Oberlech, Austria. 1986.
69. Demay, Y., Maurel, P., Gourrier, S.: Interaction of Si(111) surface with H_2, NH_3, SiH_4 multipolar plasmas studied by *in situ* ellipsometry. J. de Phys., <u>C10</u>, 253 (1983).
70. Kinosita, K., Nishibori, M.: Porosity of MgF2 films – evaluation based on changes in refractive index due to adsorption of vapours. J. Vac. Sci. Technol., <u>6</u>, 730 (1969).
71. Lichtenecker, N.: Dielektrizitätskonstante von Mischkörpern. Phys. Zeitschr., <u>27</u>, 115 (1926).
72. Juran, D.: Mikrocomputergesteuerte Motor–Geschwindigkeitsregelung. Vienna: Techn. Univ. Vienna, Diploma Thesis, in preparation.
73. Müller, W.: Photodetektor–Vorverstärker. Vienna: Techn. Univ. Vienna, Diploma Thesis, in preparation.
74. Intel Corporation: iRMX 80 user's guide. Santa Clara, Calif.: Intel Corp. 1980.
75. Intel Corporation: ISIS–II user's guide. Santa Clara, Calif.: Intel Corp. 1983.

Subject Index